Changing Prairie Landscapes

Changing Prairie Landscapes

Edited by

Todd A. Radenbaugh
and
Patrick C. Douaud

Canadian Plains Research Center
University of Regina
2000

Canadian Plains Research Center
University of Regina
Regina, Saskatchewan S4S 0A2

Canadian Cataloguing in Publication Data

Main entry under title:

Changing Prairie landscapes

   (Canadian plains proceedings, ISSN 0317-6401 ; 32)

Articles originating in the "Plain as the Eye Can See" Conference, held in Regina, April 15-16, 2000.

   Includes bibliographical references.

   ISBN 0-88977-146-4

1. Prairie Provinces — Geography — Congresses. 2. Human ecology — Prairie Provinces — Congresses. 3. Nature — Effect of human beings on — Prairie Provinces — Congresses. I. Douaud, Patrick C., 1949– II. Radenbaugh, Todd, 1964– III. Plain as the Eye Can See Conference (2000 : Regina). IV. University of Regina. Canadian Plains Research Center. V. Series.

GF512.P7 C47 2000   304.2'09712   C00-920226-9

Cover design by: Brian Wood, Brian Wood Design Studio, Regina, Saskatchewan

Printed and bound in Canada by
Houghton Boston, Saskatoon, Saskatchewan
Printed on acid-free paper

CONTENTS

INTRODUCTION

CHANGING LANDSCAPES OF THE NORTHERN GREAT PLAINS

Todd A. Radenbaugh and Patrick C. Douaud

Landscapes

Oxford Dictionary 2nd Edition 1989

Landscape (n),

1.a. A picture representing natural inland scenery, as distinguished from a sea picture, a portrait, etc.
   b. The background of scenery in a portrait or figure-painting.

2.a. A view or prospect of natural inland scenery, such as can be taken in at a glance from one point of view; a piece of country scenery.
   b. A tract of land with its distinguishing characteristics and features, esp. considered as a product of modifying or shaping processes and agents (usually natural).

3. In generalized sense (from 1 and 2): Inland natural scenery, or its representation in painting.

4. In various transf. and fig.uses.
   a. A view, prospect of something.
   b. A distant prospect: a vista. (Cf. 2b.)
   c. The object of one's gaze.
   d. A sketch, adumbration, outline; occas. a faint or shadowy representation.
   e. A compendium, epitome.
   f. A bird's-eye view; a plan, sketch, map.
   g. The depiction or description of something in words.
   h. Other transf. and fig. uses.

Landscape (v),

1. To represent as a landscape; to picture, depict.

2. To lay out (a garden, etc.) as a landscape; to conceal or embellish (a building, road, etc.) by making it part of a continuous and harmonious landscape.

The concept of landscape has a wide base and is inherently interdisciplinary, having different significance and containing multiple definitions that depend entirely upon the perspective of the user. Landscapes can be a collection of geological landforms, framed around an ecological region such as a prairie, an activity like farming, or a mental and social construct as in railroads. They can be viewed from single panoramic viewpoints taking in the broad "bird's-eye-view" or be multidimensional, containing facets of preconceived notions coupled with the physical realities. From a society's perspective, landscapes identify the general appearance of the land via the necessary institutions, political organizations and infrastructures. There often is a dimension of "improvement" to the native state by human

1

technology, as illustrated by the terms "landscaping" and "landscape architecture." On the other hand, landscape ecology is concerned with the development and dynamics of the spatial heterogeneity of an area on a broad scale, and with its effects on biotic and abiotic processes.

Here, then, landscapes are viewed in terms of biological, physical, and social heterogeneity. Landscapes are composed of many identifiable and relatively homogeneous patches, made up of ecological (e.g. habitats or species assemblages), physical (e.g. landforms such as streams, watersheds, and depositional basins) or social (e.g. farmyards, towns, cities, etc) units. Further, landscapes and their components can vary in spatial extent. The recognition of a particular area as either a patch or a landscape is a function of scale and of the system being investigated: for example, what is a single patch of grass to a browsing bison remains a complex landscape to an ant colony. Further, landscapes have a temporal component and are rarely static: rather, they are constantly changing and evolving — either in the mind or through ecological, geological, and social processes. Further, the same piece of land might be perceived differently when taking into account history, be it ecological, geological, or social. Estimating or reminiscing on the way it used to be when the bison grazed there, glaciers were melting, or when passenger trains brought in new settlers, all conjure up different historical landscapes for a region. For these reasons, landscapes are clearly hierarchical, with many fluid spatio-temporal units at fine-scale being embedded within an evolving larger system. Thus, landscapes here are seen as mosaics of ecological, geological, and social patches whose number, size, and spacing are proportional to the perspective of the viewer.

Although change is inevitable, human thinking has traditionally preferred to see *rest*, however defined, as the ultimate goal of any evolution, be it geological or cultural. For example, the geographer Reynaud once stated: "Christianity yearns for rest, i.e. the Last Judgement; Marxism yearns for rest, i.e. the establishment of global communism, which will put an end to class struggle; and geomorphology yearns for rest, i.e. the peneplain" (Reynaud 1971: 76). Yet, even this early notion of peneplain and plain (landforms molded by erosion and to a lesser extent deposition) has more recently been shown to be anything but static, and therefore replaced by the concept of a dynamic landscape equilibrium. Thus, there seems to be an innate yearning in humans to perceive stability in, or physically force stasis on, landscapes. But the realities of landscapes are far from this notion. Landscapes and the systems that form them change — despite, or many times owing to, human efforts to maintain stability. The present collection of articles explores many of these multiple dimensions and definitions concerning changing landscapes within a specific region.

Landscapes of the Northern Great Plains

In common with landscapes everywhere, those on the northern Great Plains are also evolving and changing. One example of drastic change on the prairies is seen in the ecological landscape. As it stands, approximately 80% of the northern Great Plains have been converted from native habitats into other land cover types (Selby and Santry 1996; Samson and Knopf 1996). Moreover, since the late 1880s the two dominant structuring forces, fire (Rowe 1969; Wright and Bailey 1980; Collins and Wallace 1990; Collins 1992) and large grazing ungulates (Frank et al. 1998; Knapp et al. 1999), have been eliminated or heavily managed by primarily agricultural activities. Meanwhile, social institutions have not remained static either: since the mid-1930s there has been a substantial decline in the rural population, which has caused many alterations in this landscape despite heavy resistance to this inevitable change. The magnitude of these and other forces raises concerns about the possible consequences of human-induced change on the prairie landscape.

The historical context of the northern Great Plains also gives insight into how we have come to perceive landscapes. Prairie pioneers in letters back home generally did not describe the biotic landscape in great detail, but often expounded on extremes of weather and climate, and how these affected social landscapes. In Saskatchewan of the 1920s and 1930s, Jane Aberson wrote many articles for a Dutch newspaper about her experiences establishing a homestead. In several of these articles the weather took on great importance; for instance in April 1934 she wrote:

> It's the second week of April and we take it more or less for granted that spring is just around the corner. But March stung us with temperatures thirty degrees below zero. I feel downright miserable when I think of the first violets and blossoms on bushes that Hollanders see in March. But even lifelong Canadians start to grumble about stinging cold weather in March. 'The longest winter I have ever seen' is the standard comment of practically every old-timer I've seen in the past few weeks. But to be quite honest we hear that all the time: the hottest summer, the driest spring, the coldest spring, the wettest fall. I start to think that their memories are not any too long. (Aberson 1991: 116)

Further, early settlers typically described their civic contribution to the region or their painstaking efforts to "improve" on raw nature. For example, the geologist and explorer H.Y. Hind declared after touring the Canadian Prairies in 1857 and 1858 that:

> It is a physical reality of the highest importance to the interest of British North America that this continuous belt [aspen parkland and moist mixed grassland] can be settled and cultivated from a few miles west of the Lake of the Woods to the passes of the Rocky Mountains, and any line of communication, whether by waggon [sic] road or railroad, passing through it, will eventually enjoy the great advantage of being fed by an agricultural population from one extremity to another. (in Friesen 1984: 302)

Even the naturalists Coules and Macoun, who (in the 1870s and late 1890s respectively) surveyed the northern Great Plains at the beginning of settlement, were often transparent in their biases concerning the local potential for European habitation. Macoun even went as far as claiming the arid Palliser's Triangle as "well suited for agriculture" (in Friesen 1984: 179). This Canadian Prairie region in the early 1900s was known as the "last best West" — referring to the viewpoint that the region was the last opportunity to obtain free homesteads in North America.

Within the whole northern Great Plains, early settlers constructed an image of a future landscape where existence and prosperity were seen only in the context of an expanding European society. In a contemporary sense we are living examples of this legacy. However, the society we have built is still intermingled with the natural systems in which it is embedded, marking landscapes as human constructs as well as physical spaces.

As our technology, population and affluence have increased, prairie landscapes have changed ever more rapidly. Never before has society had more influence on them as it does now. Even as scientists, humanists, policy makers, and many others argue the etiology, extent, and rate of human influences in altering landscapes, change to them is occurring. Granted, earlier societies and other species may also have caused change; but the rate, level and scope of the change occurring today are unprecedented. Moreover, this change is multifaceted, having an impact not only on the fabric of culture and its perception of landscape, but also on the ecology (i.e. changes in prairie biodiversity and species interactions) and physical landforms (i.e. mining eskers for gravel or ancient swamps for coal), both of which are interlinked with human society. Over the past decade, scientists from many disciplines have therefore paid increased attention to the long-term role society has had in structuring the patterns and processes of prairie landscapes. For example, much effort has been put in identifying and protecting plants and animals that have become rare or endangered and, more recently, in the conservation of habitats that harbour these species.

Interdisciplinary and multidisciplinary research is therefore becoming an important tool in identifying the influences that human activities have not only on cultural landscapes but on biophysical ones as well. The strength of this research comes from the integration of disciplines focusing on broad-scale questions. The need for this integration has been recognized by many organizations and institutions. Such interdisciplinary approaches as the study of Chesapeake Bay (Simpson and Christensen 1997) and Canada's Prairie Conservation Action Plan (Dyson 1996) have already led to major advances in our understanding of how ecosystems work in conjunction with our social and political systems. This collection of articles focuses on just such an integration in the northern Great Plains of North America.

The Prairie Play?

There is a popular analogy to the effect that nature works as an "ecological play" (e.g. Hutchinson 1965). This play occurs on a theatrical set that is constantly being constructed by physical and biological processes. The various species are the actors and set builders, and their interactions and functional roles write the script. Some species have disproportionate roles such as keystone species, dominant herbivores, and systems engineers (Lawton 1987). The most dominant species may act as the director, controlling how the play is performed, or enact leading roles in which the other actors take minor roles.

Within the past few hundred years, it has become increasingly clear to workers in such disparate fields as the geological, biological, archaeological and social sciences that in the context of the ecological play, society (and the culture that constructs it) is becoming at once director, stage engineer and leading actor. Physical, chemical, and biological processes of the Earth can therefore no longer be studied outside the sphere of human activity, and conversely society cannot be studied without taking into account the biosphere. The relationship between society and the natural landscape led to philosophical speculations such as those of Kant, who in the 18th century became fascinated by human diversity and power of adaptation — "an adaptation that presupposes not only man's faculties but a certain responsiveness or even cooperation of nature itself" (Yovel 1980: 128). This symbiotic relationship, only recently rediscovered in Western scientific thinking, has long been a spiritual mainstay of non-industrial societies. However, the contribution of biophysical processes has not been lost on anthropologists like Keith Basso, who in his analysis of the Apache perception of landscape states:

> I would like to witness the development of a cultural ecology that is cultural in the fullest sense, a broader and more flexible approach to the study of man-land relationships in which the symbolic properties of environmental phenomena receive the same kind of attention that has traditionally been given to their material counterparts. (Basso 1996: 67–68)

Once recognizing this reality, the relationship between society, ecology and geology can be explored in more detail. The actions of society have recently been shown to influence, interact with, alter, and even control the operations of biophysical functions at many levels of biological organization. Studies have shown that within many landscapes, human activities have dominant roles, directly controlling the entire system. This includes land-use practices that alter native habitat and expatriate species (Kaiser and Gallagher 1997, Tilman et al. 1997, Vitousek et al. 1997). Examples of direct human alteration of natural landscapes and ecosystems include: desertification of marginal lands (Schlesinger et al. 1990, Mainguet 1994); eutrophication of boreal forest lakes (Schindler 1998) and coastal areas (Lapointe and O'Connell 1989); loss of coral reefs in the Caribbean (Hughes 1994); loss of soil organic matter due to agriculture in the Great Plains (Seastedt 1995); and collapse of marine fisheries (Botsford et al. 1997). In addition, human infuences are often indirect. Indirect effects are a secondary result of direct species interactions, and may alter species composition and community structure (Rosemond 1996); they include changes in regional species composition, as well as the addition of pollutants to the soil, surface and ground water, and atmosphere. Recent studies have shown that both direct and indirect effects have considerable impacts on the functioning of landscapes, and consequently on human health (Botsford et al. 1997, McMichael 1997). Because of the potential for humans to alter biotic systems through both direct and indirect influences, questions arise concerning the extent to which human-controlled processes alter the broad-level functions within landscapes.

Changes in Prairie Landscapes

This collection of articles once again confirms the old adage that nothing changes but change itself — no matter what time scale is investigated. Firstly, Binda and Nambudiri set the stage for the investigation of changing prairie landscapes by outlining the large range of possible biophysical landscapes that have existed in the northern Great Plains in the geological past. Although in geological terms society is but a recent interloper on the prairies, it has been part of the landscape since the last ice age. The date when the first settlers arrived in the region is uncertain, but people of the Clovis culture are known to have roamed from Kansas to Manitoba between 11,500 and 10,000 years ago. In this context, Boyd analyzes the remnants of some of these earlier societies, as well as the changes in the biophysical landscapes that supported them. His study illustrates how adaptable societies have been when confronted with changing landscapes.

With the completion of the transcontinental railways and the arrival of European settlers, agriculture became a much more dominant part of the landscape. Investigating the changes brought about by this event, Hopkins and Running link soil types in the Sand Hills of North Dakota to agricultural land use by comparing historic (3,000 year old) prairie vegetation with that of today. They conclude that the biota of this region is resilient to many natural stresses, but when converted into crops the landscape degrades quickly and severely. Peltzer expands on this concept by looking at the changing ecosystem functions: as native grassland is planted into tame pastures the number of ecosystem services provided to society decreases. He concludes that tame pastures offer only a fraction of the services provided by native grasslands. Smith and Radenbaugh further examine this by looking at changes in breeding bird populations over the last 150 years. The striking changes to landscapes they discuss not only concern the region's avifauna — they may also have triggered a cascading of effects to many other plant and animal groups.

With regard to the sociopolitical landscape, Harrison discusses the changing faces of populism in Alberta; this study illustrates how rapidly political pressures and moods can change, and shows the resulting impact on policy and social landscapes. Another example of this is given by Olfert and Stabler's essay on rural infrastructure in Saskatchewan. Technological advances in transportation and agriculture, coupled with rural depopulation, have caused the rural landscapes to witness one of the most dramatic alterations ever wrought upon them. Further, although social resistance to this change is strong many feel that it is inevitable. Another good example of such passionate resistance to change in the landscape concerns the infrastructure that helped made agrosystems possible. Railroads were instrumental in determining patterns of European settlement and have played an important role in the social structure of the region for the past 100 years. With this in mind, Paul discusses the changing landscapes of railroads, along with the presently realities that govern the economic climate responsible for branch-line abandonment and construction of new main lines.

Our preconceived notions of a region are often dominated by weather and climate, but the realities of the physical landscape can often be overpowered by a wishful cultural landscape. Thus, people can be lured by false promises. This aspect is investigated in Kansas by DeBres, who shows how settlement patterns and a regional ethos often develop from preconceived notions of drought vs. agricultural bounty. Many early settlers and specialists asserted that the lack of moisture on the Plains would change with settlement; this claim assured homesteaders of abundant crops since the argument mentioned that rainfall *will* increase westward as sod is broken. Through the collection of accurate climatological data for the United States, the Smithsonian Meteorological Project in the mid-1800s did much to dispute or confirm these claims. Despite these data, however, many of the old perceptions of prairie climate persisted and often could even be seen in the biases affecting the Smithsonian data collectors.

Lastly, Boehm et al. examine the possible landscapes that either sequester or produce greenhouse gases by looking at the contributions modern prairie agriculture makes to total Canadian output. They investigate various scenarios of land use, and suggest ways to globally reduce greenhouse gas build-up by sequestering carbon in the soil locally. Given the business as usual scenario, the future will see the region continue to rely heavily on fossil fuels producing more greenhouse gases and therefore adding to global warming. However, many alternative scenarios, if adopted by society, could mitigate these emissions, thereby lowering the greenhouse gas output of the region.

Conclusion

Let us now return to the analogy of the play and ask ourselves whether it is an accurate one. Can the activities of the actors on prairie landscapes really be considered a play? A play is highly structured, produced through the writing of a script, the casting of actors, and the building of a fixed set, while all is overseen by a stage director. In a play, although there may be differences between successive performances, or even alternative productions, no show ever changes substantially from the script. At the end of *Oklahoma*, for instance, the likable Curly McLain will always win Laurey Williams' hand at the box social; and the musical will end with the sinister Jud Fry falling on his own knife. However, nature does not appear to work this way, for circumstances could arise that may deviate from the script. Gould (1989) argues convincingly that the history of life has too many contingencies for multiple productions to have the same ending. For example, in nature, if you could rewind the tape and play it back, there might be a new version where Laurey and Curly never fell in love or even met. Jud could even win the fight after alighting the haystack, turning the play from a happy love story into a tragedy.

The adventures of Laura Ingalls Wilder and her family in their journey west by covered wagon (told in her numerous *Little House on the Prairie* books) further illustrate the importance of contingency. Born in 1867 in a Wisconsin log cabin, Laura soon migrated to Kansas with her family, only to find they were in Indian Territory; they thus were forced to move on. In their travels, strife and hardships became the rule, from Grasshopper Country in Minnesota to a sister going blind (Zochert 1976). One feels that they could have been happy settling in many of the locations they traveled, which included Wisconsin, Kansas, Minnesota, Iowa, South Dakota, Florida, and finally Missouri. However, uncontrollable factors always intervened which compelled them to respond to forces in the landscape, each time producing different results and adventures. Given a different set of events, or a different time period, the narrative may have told of the family taking the Oregon Trail to California or perhaps joining the land rush to the Canadian Prairies in 1910.

If this is the case, then improvisational theatre or improvisational story writing may provide a more appropriate analogy. Here the story and characters are highly contingent, with the course of the action changing with every performance or composition. Although there is no script or single author, acts or chapters are not chaotic but can be highly entertaining. Moreover, there are rules that structure and control the action, but they are only roughly sketched out by the players or writers. If people enter during the middle, they are seldom lost and will soon discover the rules simply by observing the action, even if these rules change.

Interdisciplinary research can be seen as this late comer. Since it tends to view landscapes from a broad and multiple perspective, researchers taking such an approach may have a better chance to see things that specialists might miss. This allows new insight into the patterns and processes that create and change landscapes. Thus, after observing the show we, as interdisciplinary researchers, have an opportunity to get together and figure out the rules. This is precisely what this collection of articles is trying to achieve.

Acknowledgements

We would like to thank Randy Widdis, Alec Paul, Dave Sauchyn, Brian Mlazgar, David Miller and Wendee Kubik for their helpful suggestions and comments. We also thank Simone Hengen for our discussions on contingency.

CHAPTER 1

CHANGES IN LANDSCAPE AND CLIMATE IN THE CANADIAN PRAIRIES AT THE END OF THE MESOZOIC ERA

Pier L. Binda and *E.M. Vasu Nambudiri*

ABSTRACT. This article discusses changes in landscape that occurred over 65 million years ago. The Late Cretaceous–Paleocene sedimentary succession of the Canadian Prairies records changes in landscape and climate. An inland sea intermittently occupied most of the three prairie provinces throughout the Cretaceous Period (146 to 65 Ma). During Maastrichtian time (70 to 65 Ma), as the sea retreated to the south, rivers flowing from the rising mountains in the west shed sediments onto the plains where swamps and lakes developed. A lush, tropical-to-warm-temperate vegetation contributed to the development of numerous coal seams. Subtle climatic variations can be detected from changes in pollen and spore assemblages. A gradual climatic cooling occurred in the last two million years of the Cretaceous Period and continued into the Paleocene Epoch (65 to 62 Ma). The impact event of the terminal Cretaceous, which is interpreted by some as the cause of mass extinction, seems to have had little effect on the vegetation of the region.

Introduction

In the last three decades of the millennium, climate changes both at the global and at the local scale have received a great deal of attention, mainly because of what some scientists call "the anthropogenic forcing" (Fyfe 1992). Unfortunately, most of the debate on global warming and the projected effects on the planet has taken place without considering the great changes in climatic conditions that occurred at the geological scale.

Whereas the eastern and western seaboards of North America were subjected to episodes of intense deformation and mountain building as exotic *terranes* were accreted onto it, the site of the present prairie region enjoyed a long period of relative tectonic stability. The cratonic interior of North America underwent mainly vertical movement as the whole continental plate was moving towards higher latitudes throughout approximately 500 million years (Ma) of geological history. Thus, this region is an ideal locale to investigate landscape and climate changes at the geological scale as the Prairies contain, in the subsurface and in scant outcrop, an almost 500 Ma uninterrupted record of sedimentation, testifying to frequent changes from marine to continental conditions.

In brief, the geological and geophysical record shows that during early to middle

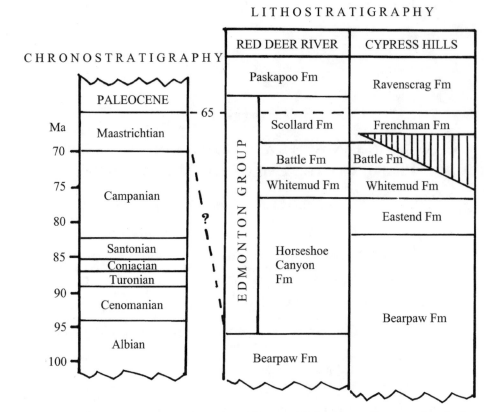

Figure 1. Late Cretaceous stratigraphy in southern Alberta and Saskatchewan. Chronostratigraphy from Obradovich and Cobban (1975); lithostratigraphy from Nambudiri and Binda (1991).

Paleozoic, the region lay at peri-equatorial latitudes (Kent 1994); shallow hypersaline seas dotted with reefs were the dominant feature of this time. The coastal areas of the Arabian peninsula may be considered an approximate modern analogue.

By Mesozoic times the region had moved to higher latitudes and, particularly in the later part of the era, great changes in geomorphology, flora and fauna were taking place. Finally, the glaciations and ice melting of the last two million years resulted in today's landscape.

In this article we focus on the relatively short time span (approximately 20 Ma) from Late Cretaceous to Early Paleocene (Figure 1), in which the region went from being covered by a long and narrow inland sea with tropical climate to a mostly warm-temperate, swampy lowland. The abundance of sedimentological, isotopic and paleontological data available allows us to make reasonably accurate paleo-climatic interpretations.

The Cretaceous Period (146 to 65 Ma)

The Cretaceous is characterized by important events affecting lithosphere, hydrosphere, biosphere, and atmosphere. At the beginning of the period, the continental plates that constituted the supercontinent Pangea were still fairly closely clustered, as only the southern continent of Gondwana had separated from

Laurasia to the north. Therefore the throughgoing tropical ocean Tethys was probably the major factor influencing climates. By the end of the Mesozoic Era, a considerable separation of North America and Europe forming the Atlantic Ocean had been achieved, allowing polar currents to flow southward and exercise an influence on later (Cenozoic) climates (Matthew 1984). Global sea-level curves based on seismic stratigraphy indicate that throughout the Cretaceous Period the sea stood higher than at present (Hallam 1981). Most geologists attribute this to high rates of sea-floor spreading and greater production of basalt at oceanic ridges.

Climatically, the Cretaceous has been considered ice-free, a "greenhouse" world with equable, warm climates (Hallam1993). Reports of ice-rafted "dropstones" from Early Cretaceous rocks in Australia and elsewhere have recently been discounted as large clasts rafted to sea by driftwood and by floating vegetation (Bennett and Doyle 1996). However, towards the end of the period, a climatic deterioration has been recorded from many parts of the world (Hallam 1981 and references therein).

One of the most fascinating episodes in the Earth's vegetational history also occurred in the Cretaceous: the sudden appearance of flowering plants. They appeared in the Early Cretaceous as a low-diversity group but rapidly expanded to attain their current status as the most dominant component of the modern flora, replacing the giant conifers of the Jurassic forests. Today, flowering plants inhabit all parts of the world and are the most successful plant colonizers in all ecological zones and climates. Their sudden appearance in Early Cretaceous prompted paleobotanists to speculate that they might have originated earlier in upland regions of the world where processes of fossilization seldom worked.

The Inland Sea (110 to 68 Ma)

From approximately 110 to 68 million years ago the site of the present-day Canadian Prairies was occupied by a narrow interior seaway. During maximum flooding the sea extended from the Arctic Ocean to the Gulf of Mexico, where it joined the vast Tethys Ocean. Caldwell (1982, 1984) summarized the history of the seaway in terms of lithostratigraphy, biostratigraphy, and of the effects that global eustatic sea-level changes and the rising of the mountains to the west had upon it. In response to transpressional forces that accreted exotic terranes to the western margin of North America, wedges of sand were intermittently spread into the basin from the west. To the east and to the northeast the sea was onlapping and offlapping Paleozoic and Precambrian landmasses. The sea was thus expanding and contracting as a result of the varying sediment input and of the global sea-level rises and falls (Haq et al. 1987).

Pollen, spores and wood trapped in carbonaceous sediments can be used to determine depositional environment, vegetation and climate along the shores of the Cretaceous continental sea. This analysis shows that the environment was roughly equivalent to the modern-day coastal swamps of the southeastern United States. The vegetation at approximately 100 to 92 Ma ago is known from the work of Singh (1964, 1971) and of Nambudiri et al. (1986). Figure 2 illustrates the vegetation profile from uplands to coastal regions. Its main elements are conifers, ferns, other pteridophytes, and cycads. There is little doubt that cycads and cypress (*Taxodium*-type) plants and the ferns that dominated these swamps are indicative of tropical to subtropical climate. However, it is debatable whether the conifer pollen found in these Cretaceous rocks was wind-borne from cooler highlands or whether the

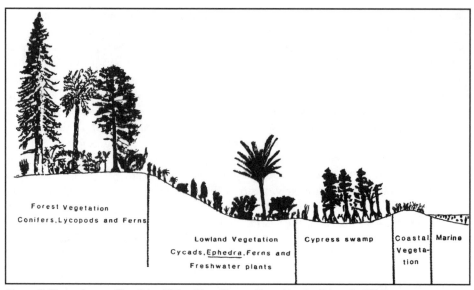

Figure 2. Vegetation profile of the Albian Manville sea (from Nambudiri et al., 1986).

conifers were able to withstand higher temperatures and humidity than their modern equivalents.

By the time of the last two expansions of the sea, approximately 85 to 70 Ma ago (Figure 3), angiosperms had become the dominant vegetational elements on land, at least in terms of pollen yield: whereas 100 Ma ago the shoreline vegetation was dominated by ferns; 20 Ma later the same ecological niche was taken over by angiosperms. On the basis of angiosperm pollen, two distinct phyto-geographic provinces, separated by the sea, can be recognized (Srivastava 1978). East of the sea, floral elements of Atlantic-European affinity constituted the Normapolles phytoprovince; to the west, the *Aquilapollenites* province extended as far as Siberia. It must be kept in mind that the provinces are named after minor, albeit characteristic components of the vegetation. In fact, *Aquilapollenites* refers to the pollen of a parasitic plant, probably a mistletoe, that lived on tupelo (*Nyssa*) and *Taxodium* trees which were some of the largest and most conspicuous components of the tropical/subtropical coal swamps.

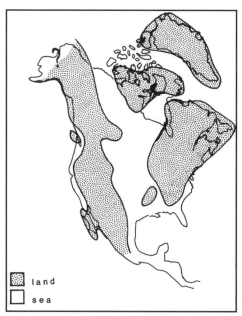

Figure 3. Maximum extent of the Bearpaw sea (after Srivastava, 1978).

Oxygen isotopic ratios from shells of ammonites living in the sea in which the Bearpaw Formation (Figure 1) was deposited approximately 70 to 68 Ma ago, suggest surface water temperatures around 20°C (see Caldwell 1982): this is warm considering that the region lay at approximately the same latitude as today, and yet it is somewhat cooler than temperatures measured by the same technique for oceans of 100 Ma ago (Hallam 1981 and references therein). Throughout the 110 to 68 Ma time span, large terrestrial reptiles lived on dry land and in the swamps, and marine reptiles dominated the sea.

Deltas, Rivers and Coal Swamps (68 to 67 Ma)

The Late Cretaceous sedimentary rocks overlying the marine shale of the Bearpaw Formation (or its equivalents) have been subdivided into a number of stratigraphic units bearing different names from the Rocky Mountains Foothills to the Alberta plains, the Cypress Hills, and Manitoba. A detailed treatment of these units is beyond the scope of this article: reference will be made only to the units exposed in the badlands of Alberta, the Cypress Hills, and south-central Saskatchewan (Figure 1). The transition between marine and continental rocks can be seen near the base of the hoodoos of Drumheller: in fact, the transition between marine and continental rocks can occur over as much as a 300m-thick section, as shown in a micropaleontological study by Wall et al. (1971) on the ratio of fern megaspores versus foraminifers.

The lower part of the section, the continental Horseshoe Canyon Formation (Plate 1), consists of approximately 250 m of sandstone, siltstone and shale, and of at least a dozen coal seams, several of which have been commercially exploited in the past (Allan and Sanderson, 1945). Detailed sedimentological studies in the Drumheller badlands indicate that as the sea retreated towards the south, a tide-dominated delta prograded over the marine sediments (Rahmani 1981). Landwards of the delta, meandering streams and swamps developed. The continental deposition was briefly interrupted by a short-lived marine incursion that deposited an oyster-rich bed (Drumheller Marine Tongue). A diverse dinosaurian fauna lived in the swamps and on the dry land. Remains of herbivorous dinosaurs such as duckbills, ceratopsians and ankylosaurs as well as carnivorous therapods and egg predator ostrich-like dinosaurs occur in the Horseshoe Canyon Formation of Alberta (D. Russell 1984).

The vertical succession of the Horseshoe Canyon Formation in the Red Deer River valley of Alberta has been subdivided into six floral zones based on pollen and spore assemblages (Srivastava 1970). The basal transition beds indicate the establishment of subtropical salt marshes inhabited mostly by halophytic plants. Above the transition beds, subtropical woodlands and rain forests sheltering fresh water marshes, ponds and brooks developed as indicated by the pollen and spores of Srivastava's (1970) zones II and III. Somewhat more humid conditions followed, as suggested by the development of cypress swamps dominated by Taxodium-type trees (zone IV). The swamps were then flooded by the short-lived marine incursion mentioned above (Drumheller Marine Tongue). After the final withdrawal of the sea, a slight cooling of the climate is indicated by the appearance of vegetation of the warm-temperate zone. Srivastava (1970) suggested that this may have been caused by elevation of the borderland after the marine regression.

A conspicuously white stratigraphic unit 10 to 20 m thick consisting of sandstone, siltstone and shale, with a thin carbonaceous layer occurs above the

Figure 4. Reconstruction of the Maastrichtian Whitemid flora of Alberta and Saskatchewan based on plant megafossils, seed cuticles, pollen and spores. The Whitemud landscape was dominated by meandering streams, lakes and cypress swamps. Dominant plant species, numerically represented in the figure are: 1) *Ginkgo biloba*, 2) *Araucaria*, 3) unknown conifers, 4) *Phyllanthus emblica*, 5) *Cercidiphyllum*, 6) *Osmunda*, 7) *Nelumbium*, 8) *Nymphaea*, 9) *Azolla*, 10) *Pandanus*, 11) *Nyssa*, 12) *Loranthaceae*, 13) *Taxodium*, 14) *Typha*, 15) lycopods, 16) *Butomus*, 17) *Juncus*, 18) Cycads and 19) *Alnus* (not to scale). (from Binda et al. 1991).

Horseshoe Canyon Formation. Known as the Whitemud Formation, it has been commercially exploited for its kaolin content in the past, particularly in south-central Saskatchewan and in the Cypress Hills, where for years it has been the raw material for the Medicine Hat potters. A few quarries are still intermittently worked, and the white unit can be seen from far away in the prairie landscape. Sedimentological and paleontological aspects of the Whitemud are treated in some detail by Binda et al. (1991) and by Nambudiri and Binda (1991). A pictorial reconstruction of its 67-Ma old floral assemblage is shown in Figure 4; the landscape was dominated by meandering streams, floodplains, and cypress swamps. The climatic cooling already detectable in the uppermost part of the Horseshoe Canyon Formation is emphasized in the Whitemud by the abundance of pollen attributed to ancestors of the modern-day birch (*Betula*), alder (*Alnus*), and hornbeam (*Carpinus*) (Srivastava 1981). However, plants of the temperate zone occur in the Whitemud strata in association with floral elements of the subtropical zone. The presence of conifer pollen, and the frequent finding by the present authors of silicified fragments of conifer wood in the Whitemud Formation of south-central Saskatchewan, corroborate the interpretation of the climatic cooling.

In spite of extensive search in the Whitemud Formation, at times involving a large number of students, we have never found vertebrate fossil remains, and none are known from the literature. And yet, at places, fossil feces (coprolites) attributable to large land reptiles (Schmitz and Binda 1991) occur in abundance (Plate 2), together with petrified fruits, some of which bear a striking resemblance to the modern-day East Indian gooseberry (Nambudiri and Binda 1989). The most likely explanation is that, after deposition of the sediment, bones and teeth that may have been present were destroyed by high levels of carbonic acid in the ground water; on the other hand, the acidic conditions favoured the fossilization of feces and fruits by replacement of the organic matter with iron carbonate.

The Whitemud Formation marks an important break in dinosaur paleontology: the diverse fauna of the Horseshoe Canyon Formation seems to disappear at this level, and only a few genera have been found in the terminal Cretaceous Scollard Formation (L. Russell 1983) and Frenchmen Formation. In particular, the duck-bills and the ostrich-like dinosaurs seem to suddenly disappear at this level.

The "Battle Lake" (ca. 66.5 Ma)

Overlying the Whitemud Forma-tion, and in sharp chromatic contrast with it, is the 3 to 12 m thick chocolate-brown bentonitic mudstone of the Battle Formation (Plate 3). The unit is so remarkably continuous over a large area in the Alberta plains and in the Cypress Hills that for a long time it was thought to represent another marine incursion in the region. However, detailed micropaleontological analysis revealed that the only fossils in the mudstone were siliceous algae, parts of siliceous sponges, and silicified megaspores of aquatic plants typical of the lacus-trine or paludal realm (Binda 1992).

Tuffaceous layers and grains of volcanic quartz scattered throughout the mud-stone (Binda and Watters 1997) attest to the fact that volcanic ash fall was constant in the lake. Ritchie (1957) showed that the ash was probably coming from volcanic eruptions around Butte, Montana. Isotopic ages obtained by Folinsbee et al. (1965) and recalculated using a more precise constant (Lerbekmo pers. comm.) indicate that the ash fall occurred 66.5 Ma ago.

D. Russell (1977) depicted the "Battle Lake" as occupying an area in excess of 400,000 km^2, roughly the size of the Caspian Sea; however, as suggested by Binda (1970), it is more likely that instead of one large, continuous body of water, there might have been a number of partially interconnected lakes, swamps, and marshes. Several modern swampy or marshy areas in non-glaciated regions of the world approach the required size: the Sudd of southern Sudan and the Llanura Santafesina of Argentina are probably better modern analogues than a single large lake. The most likely explanation for the deterioration of the drainage that brought about extensive lacustrine conditions is an uplift to the east while the mountains to the west were experiencing a phase of tectonic quiescence which would account for the scarcity of coarse detritus in the mudstone of the Battle Formation.

The abundance of fossil chrysomonad algae indicates a temperate climate with possibly some seasonal water freezing (Srivastava and Binda 1984). A temperate cli-mate is corroborated by the occurrence of megaspores related to the modern genus *Isoetes*, an aquatic plant which is common in the temperate zone. The somewhat cooler conditions suggested by the fossils of the Battle Formation can perhaps be attributed to screening of solar radiation by volcanic ash in the atmosphere: the temperate climate of the 66.5 Ma time slice could then be considered as a cool peak occurring within a general cooling trend.

The Terminal Cretaceous (66 to 65 Ma)

An erosional disconformity marks the passage between Battle Formation and overlying sediments (Figure 1). Although swampy and lacustrine conditions con-tinued at places, some channelling by rivers is indicated in sediments of the Scollard and Frenchman formations. The pollen and spore assemblage suggests that the climate was warm-temperate. A rich ground flora of pteridophytes and aquatic plants were present, conifers occupied neighbouring uplands, but the number of flowering plants appears to have been reduced throughout the interval (Srivastava 1970). It must be noted that, although humid conditions prevailed in much of the region, locally, a drier regime is suggested by the occurrence of a caliche-type paleosol in southwestern Alberta (Jerzykiewicz and Sweet 1986).

However, one of the most important and most controversial events in the history of the planet is recorded in the sediments of the Prairies. At several stratigraphic intervals in the earth's history, dramatic turnovers of major groups of organisms are recognized. These mass extinctions are global in scope and affect plants and animals, both in the terrestrial and in the marine realms. Mass extinctions are recognized at the close of both the Paleozoic and the Mesozoic Eras. Considerable attention has been given to the Terminal Cretaceous Extinction, not only because of the organisms involved, but also because of its possible causes. Globally, the Cretaceous-Cenozoic (K-T) boundary mass extinction affected several animal groups both in the sea and on land. Among the invertebrates, ammonites, rudistids and some types of planktonic organisms became extinct. Among the vertebrates, the abundance of mammals was considerably reduced; but most importantly the dinosaurs (Plate 4) were completely wiped out. However, certain groups of animals (e.g., crocodiles and turtles) survived the K-T holocaust. Geologists are sharply divided into those seeking an extraterrestrial cause and those seeking a terrestrial one for the K-T event (see reviews by Sweet et al. 1999 and by Hallam 1987).

Different apocalyptic K-T boundary scenarios, possibly related to the impact of a large extraterrestrial body at Chicxulub in the Yucatán and having had worldwide effects, are reviewed by Sweet et al. (1999) who summarized the sequence of events as follows:

1. a pre-impact supernova which caused environmental perturbations possibly lasting as long as 5,000 years;
2. an impact of an extraterrestrial body followed by a very strong heat pulse and by the fallout of a blanket of ejecta which included microtektites (siliceous spherules);
3. a post-impact dust-and-aerosol-induced period of darkness possibly lasting a few years;
4. an acid rain caused by oxidation of atmospheric nitrogen and sulphur dioxide;
5. a return to full sunlight;
6. a greenhouse effect caused by abnormal concentrations of carbon dioxide;
7. a return to "normal" climate as the impact-generated carbon dioxide entered both terrestrial and marine sinks.

A fireball and extensive wildfires are suggested by various authors to have occurred as a result of the impacts, leaving as a record a typical clay layer rich in microspherules and shocked quartz, and yielding anomalously high contents of the platinum-group metal iridium.

Opponents of the impact theory (e.g. Hallam 1987) maintain that most of the above-mentioned features can be explained by volcanic activity caused by significant disturbances in the mantle. An enrichment in iridium 100,000 times higher than normal has been measured in aerosol released by one particular eruption of the Kilauea volcano in Hawaii; moreover, microspherules are not unique to the K-T boundary, and shocked minerals may occur in regions adjacent to explosive volcanic events.

Whether the global extinction of dinosaurs can be attributed to the impact of an extraterrestrial object or to a gradual dying-out related to other causes remains a geological and biological puzzle. Many theories have been put forward on the subject (Hallam 1981 and references therein). They include, among others: reduction of habitat due to low topographic diversity; high rate of magnetic reversals causing a weakened magnetic field and consequent reduced protection against the influx of harmful cosmic radiations; climatic deterioration; predation of nests and eggs by mammals; changes in diet that caused either constipation or alkaloid poisoning; allergies to pollen of flowering plants (Srivastava 1978).

On the Canadian Prairies the diversity of dinosaurs decreased considerably in the latest Cretaceous (L. Russell 1983). Furthermore, the last dinosaur bones found in the region occur a few metres below the K-T boundary (Lerbekmo et al. 1979). Had the demise of dinosaurs been nearly instantaneous as required by the impact scenario, large accumulations of bones would be expected to be found at the impact level. However, as suggested by Stanley (1998), the paleontological record and our knowledge of it can never be as complete as we would like.

The recent detailed study by Sweet et al. (1999) for the Canadian Prairies does not support the cataclysmic scenario invoked by some authors, who suggested that a temporary disappearance of flowering plants occurred at the K-T boundary marked by a great abundance of spores of opportunistic ferns. Sweet et al. (1999, p. 766) stated that "palynofloristic evidence suggests a gentler and more regionally restricted impact-induced K-T boundary environmental perturbation than the living hell." The western Canadian record indicates that throughout the K-T "event" the level of light was sufficient to allow photosynthesis and, although floristic changes did occur (possibly involving the loss of some of the forest canopy), pollen of flowering plants occurs in some abundance in the boundary claystone. Furthermore, coal swamps and a high water table persisted across the K-T boundary as witnessed by the Nevis coal seam immediately overlying the boundary claystone. This suggests high precipitation during a time in which, according to proponents of the catastrophic event, levels of CO_2 were as much as fifty times higher than at present. Sweet et al. (1999) conclude with the speculation that, if such high levels of carbon dioxide in the atmosphere did not change the precipitation patterns, projections of a future mid-continent drying out by a rapid infusion of CO_2 into the atmosphere should be reassessed in light of the latest Cretaceous and earliest Paleocene record in the Canadian Prairies.

The Paleocene Epoch (65 to 58 MA)

The Paleocene Epoch is the prelude to the adaptive radiation of many modern species. In the terrestrial realm, as is the case for the Canadian Prairies, changes in vegetation and climate were almost insignificant during the Paleocene. Gone were the large reptiles that had dominated the Mesozoic Era, but the mammals had not yet managed to take full advantage of the opportunity to occupy the available ecological niches vacated by the dinosaurs.

However, two aspects of the landscape changed slightly during this epoch: rivers flowing from the rising mountains in the west supplied coarser sediments to the plains of Alberta, and the coal swamps slowly migrated towards the southeast. This

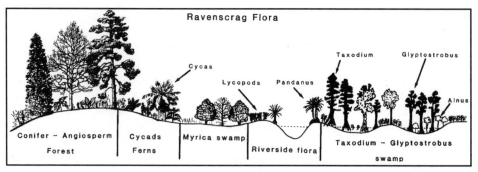

Figure 5. Vegetation profile of the Paleocene Ravenscrag Formation (from Potter et al., 1991).

is evidenced by the fact that, whereas only one minor and one major coal seam, the Nevis and the Ardley respectively, are present in Alberta, several coal seams occur in the Paleocene succession of southern Saskatchewan (Frank 1999). The southeast shift of coal swamps probably reflects a shift in the belt of high precipitation that followed the withdrawal of the sea to more southerly regions. A cross-section of vegetation zones in the Willow Bunch coal field of southern Saskatchewan is shown in Figure 5 (from Potter et al. 1991). *Taxodium* and *Glyptostrobus*, together with *Nyssa*, were the dominant large trees of the swamps whereas conifers and Betulaceae dominated upland forests.

The flora contained elements of both subtropical and temperate zones. Thus, it is safe to interpret the vegetal assemblage as indicating warm-temperate climatic conditions, perhaps slightly cooler than the terminal Cretaceous: the abundance of conifer pollen suggests it. However, as already discussed for the Early Cretaceous, it must be borne in mind that bisaccate conifer pollen can be transported by the wind a long distance from its place of origin: its presence in the sediments of the coal swamps may thus reflect the vegetation of distant uplands.

Conclusions

The cratonic interior of North America underwent profound changes in climate and landscape in the last 500 Ma of geological history. During the Late Cretaceous Epoch (approximately 94 to 65 Ma ago), in response to the rise of the Rocky Mountains in the west, sediments were being shed into the region that was later to become the Canadian Prairies.

The Late Cretaceous sediments of the Canadian Prairies and their vegetal microfossil content allow us to reconstruct the landscape and the climate of the period at a reasonable level of confidence. The withdrawal of the inland sea approximately 68 Ma ago is well documented in the badlands of Alberta and Saskatchewan; fluviatile and lacustrine conditions are validated by spores, pollen, seed cuticles, leaves, fruits, and by algae and sponge remains (Binda et al. 1991; Binda 1992). One of the most interesting among recent findings is that the "catastrophic event" of the K-T boundary does not appear to have had a great effect on the vegetation of the region (Sweet et al. 1999); but whether this negates the global effects of the Yucatán impact and therefore supports a gradualistic interpretation of the terminal Cretaceous mass extinction, remains an open question. Isotopic age determinations carry with them a degree of uncertainty that precludes a precise determination of the time of extinction. Furthermore, the fossil record is never as complete as would be desirable.

In viewing the Cretaceous-Cenozoic boundary event, what is perhaps more intriguing in terms of relevance to modern-day climatic projections is the seemingly little effect that the alleged tremendous increase in atmospheric carbon dioxide had on the vegetation and climate of the Canadian Prairies assuming that the impact scenario envisaged by various authors corresponds to past reality.

CHANGING PHYSICAL AND ECOLOGICAL LANDSCAPES IN SOUTHWESTERN MANITOBA IN RELATION TO FOLSOM (11,000–10,000 BP) AND MCKEAN (4,000–3,000 BP) SITE DISTRIBUTIONS

Matthew Boyd

ABSTRACT. Landscape and vegetation have changed dramatically across the Canadian prairies from the terminal late Pleistocene to the end of the middle Holocene. These transformations, in turn, are often assumed to have fundamentally impacted the ways in which Native Peoples have mapped on to these landscapes through time. In this study, a comparison of Folsom (11,000–10,000 BP) and McKean (4,000–3,000 BP) archaeological site distributions in the glacial Lake Hind basin of southwestern Manitoba illustrates the extent to which landscape evolution, environment, and settlement strategies are truly entangled. These site distributions are made meaningful in relation to lithostratigraphic and paleoecologic data obtained from a cutbank of the Souris River, in the Lauder Sandhills region of the Hind basin. Folsom incursions into the Hind basin correlate well with the period of gradual drainage of glacial Lake Hind, prior to c. 10,400 BP. The regional clustering of McKean sites across the basin, furthermore, correlates with the initiation of extensive eolian landscape stability and the widespread development of a prairie wetland mosaic in this area. While both physical and biotic landscapes have changed dramatically through this period of time, however, both Folsom and McKean site distributions may reflect a highly similar land-use strategy.

Introduction

Traditional Quaternary paleoenvironmental and geomorphic research has had, as a primary goal, the development of models that address regional changes in landscape form and biotic community composition (e.g., Antevs 1955; Ritchie 1976; Sorenson 1977; Knox 1983; Barnosky 1989; Lemmen 1996). While the correlation of these models to past human activities is often an explicit goal (COHMAP 1988; Wright et al. 1993), this objective has been largely neglected in practice. This neglect has the effect of perpetuating the image of pre-European landscapes in North America as entirely "natural" surfaces which are unaffected by, and without effect on, the course of human history.

This article addresses the extent to which human history, landscape evolution, and biotic change are truly intertwined through a comparison of two archaeological complexes[1] in southwestern Manitoba. These complexes — Folsom (c. 11,000–10,000 BP) and McKean (c. 4,000–3,000 BP) — have been chosen because they are the best represented archaeological materials in the study area produced during the terminal late Pleistocene (12,000–10,000 BP) and the beginning of the late Holocene (10,000–present), respectively. As a result, they provide the best means of illustrating the connection between the archaeological record and the dramatic sequence of paleoenvironmental and geomorphic changes initiated following deglaciation. Since little archaeological evidence is presently available for the period between 10,000 and 5,000 BP in southwestern Manitoba, an analysis of intermediate "cultures" (e.g., Mummy Cave Series [7,500–4,700 BP]) is not attempted.

Study Area and Modern Setting

The study area is defined as the south-central glacial Lake Hind basin, located in southwestern Manitoba (Figure 1). The modern landscape is a discontinuous dune field (the Lauder Sandhills) which stretches for approximately 25 km on the north side of the Souris River, between the towns of Hartney and Melita. Some of the most striking eolian landforms in the study area are large, stabilized, crescentic, parabolic sand dunes (6–10 m high, >500–1500 m long) which survive mostly as northwest-southeast trending parallel ridges. Other landforms include low conical, irregular, or sinuous mound dunes (1–3 m high, 4–0 m in diameter with no slipfaces), and eolian sand sheets (ranging from 1–3 m in thickness) in interdunal areas. Interdunal swales in the sand sheets are commonly occupied by shallow wetlands. Soils in the study area are mostly Orthic Regosols (Eilers et al. 1978; Soil Classification Working Group 1998: 117), with gleyed soils frequently occupying lower topographic positions. Prior to the initiation of a land drainage program in 1969, small wetlands were extremely numerous across the study area — a fact well illustrated by the nineteenth century Dominion Land Survey maps (Hamilton and Nicholson 1999). The proliferation of wetlands in the Lauder Sandhills stems from the underlying presence of the Oak Lake aquifer, which occupies most of the former glacial Lake Hind basin (Hamilton and Nicholson 1999) (Figure 2). This aquifer was formed by the entrapment of groundwater above impervious sedimentary and bedrock deposits (Hamilton and Nicholson 1999; Boyd 2000) (Figure 2). Throughout the Holocene, groundwater accumulated within the overlying sandy deposits to the point that surface undulations filled to create small lakes and sloughs (Hamilton and Nicholson 1999: 9). The development of the Oak Lake aquifer is a significant aspect of the middle Holocene geological record in the study area, and will be given further analysis in a later section.

An extremely diverse plant association has developed within this variable topography (Plate 5). Modern vegetation surveys (e.g., Hohn and Parsons 1993) identify at least five plant communities within the Lauder Sandhills:

1) *Aspen forest*: dominated by *Populus tremuloides* Michx. and *P. balsamifera* L. with *Quercus macrocarpa* Michx. on small sand ridges, and common parkland associates in the understory such as *Symphoricarpos occidentalis* Hook., *Prunus virginiana* L., and *Amelanchier alnifolia* Nutt.

2) *Forest-grassland transitional areas*: incorporate a higher frequency of grasses

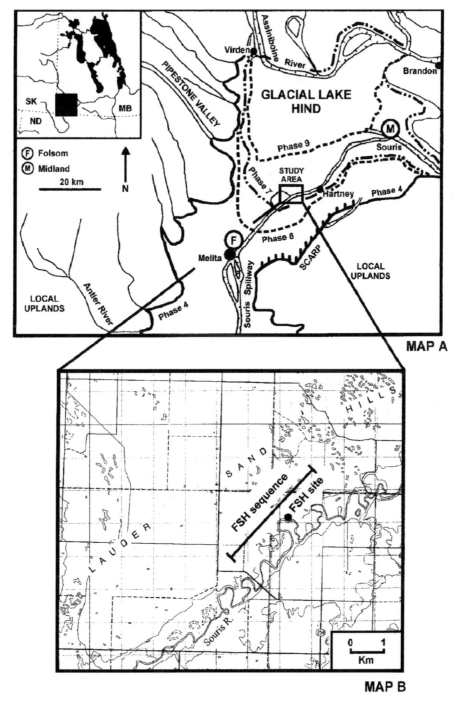

Figure 1. Paleogeography of glacial Lake Hind and location of study area. Map A: Glacial Lake Hind basin. Phase 4 to Phase 9 shorelines after Sun (1996). Map B: Close-up of study area with Flintstone Hill (FSH) site and minimum distance of FSH lithostratigraphic sequence.

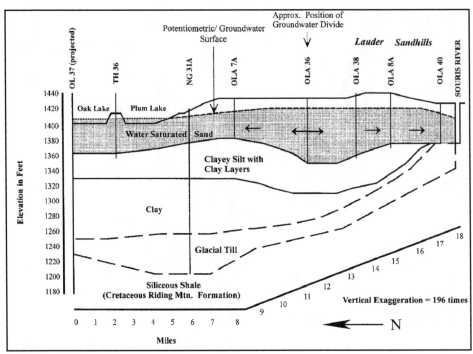

Figure 2. Sedimentological cross-section of the Oak Lake Aquifer. After Hamilton and Nicholson (1999, Figure 6) and Manitoba Department of Natural Resources Water Resources Branch File #90-1-7-1141, 1986.

(compared to the aspen forest). These grasses are predominantly subfamily Festucoideae taxa such as *Poa* spp. and *Calamovilfa longifolia* (Hook.) Scribn., and are associated with open *P. tremuloides* forest and common parkland associates.

3) *Grassland*: this rapidly diminishing habitat includes a range of prairie plant species distributed according to topography (e.g., slope position and aspect) and moisture regime. In the remnant grasslands, ground cover is dominated by *Ambrosia psilostachya* DC., *Rosa arkansana* Porter, *Equisetum hyemale* L. var. *affine* (Engelm.) A.A. Eaton, and *Solidago* spp. Native grasses in these regions are dominated by species in the Chloridoideae (warm, dry short grasses, e.g., *Bouteloua gracilis* [HBK.] Lag.) and Panicoideae (warm, moist tall grasses, e.g., *Andropogon gerardii* F. Vitman subsp. *hallii* [Hack.] J. Wipff [Wipff 1996]) sub-families.

4) *Sandhills proper*: dominated by Quercus macrocarpa stands on the northern dune slopes, interspersed with *Populus tremuloides*, diverse grass species, *Juniperus horizontalis* Moench, as well as the regionally rare *Tradescantia occidentalis* (Commelinaceae) (Britt.) Smyth.

5) *Wetlands (interdunal and lacustrine)*: common emergent taxa include *Caltha palustris* L., *Petasites sagittatus* (Pursh) A. Gray, and *Typha angustifolia* L.

More complete plant catalogues for this unique area are contained in several sources (Scoggan, 1953; Hohn and Parsons 1993; Boyd 2000).

Normal climate values at the nearest station (Deloraine) include a mean maximum summer temperature (in July) of 19.7°C, with the coldest month (January) averaging about -16.3°C. Yearly rainfall is approximately 360 mm (114 mm falling as snow), and the current ratio of precipitation to potential evapotranspiration (P/PE) is about 0.8 (Environment Canada, 1993). In general, the study area is within the subhumid continental climate zone ("Dfb" under the Koeppen system) (Eilers et al. 1978).

Archaeological and Geological Sources Used in This Study

The Archaeological Record

Archaeological research in the Lauder Sandhills has been largely confined to the "Makotchi-Ded Dontipi" locale (see Hamilton and Nicholson, 1999), a contemporary Dakota designation for a sub-region on the western end of the Sandhills. This archaeological locale defines a cluster of at least thirteen distinct sites of mostly Late Precontact (c. 2000 BP to historic period) cultural affiliation. Archaeological materials dating to the early and middle Holocene, however, consist almost entirely of "culturally" diagnostic projectile points recovered as surface finds, rather than from undisturbed and buried sites. This study draws largely from professional archaeological surveys, complemented by excavated site reports where available.

The most extensive professional surface collection program in Manitoba was undertaken as part of the Glacial Lake Agassiz Survey (e.g., Syms 1970). This database led to the production of Paleoindian (e.g., Buchner and Pettipas 1990) and McKean complex (Syms 1970) projectile point distribution maps for much of southern Manitoba. While several new sites have been located since the termination of this survey program, it remains an important source of information on regional spatial patterning of diagnostic archaeological materials across the province. Other surveys included in this study are focused on smaller areas (e.g., the Winnipeg River system [Buchner 1982]).

Three fundamental limitations are placed on interpretations derived from surface collected material. For one, archaeological materials collected in these contexts may not usually be dated by any absolute (i.e., radiometric) means. As a result, one is confined to the total known time range established for a diagnostic type across its entire geographical range. For example, the oldest Folsom date (10,900 BP) is at Owl Cave, Idaho (Miller and Dort 1978), and the youngest is from the Lubbock site, Texas, at 9,900 BP (Haynes 1964: 1410). Hence, the temporal range for any given Folsom surface recovery in southern Manitoba can only be assumed to fall somewhere between c. 11,000 and 10,000 BP. It should be noted, however, that the majority of Folsom sites in North America are usually dated to between 11,000 and 10,400 BP (Haynes 1982: 384).

The second limitation, closely related to the first, is that surface recoveries seldom provide clear and reliable spatial associations between the diagnostic projectile point(s) and other contemporaneous materials (e.g., faunal remains, other artifacts, etc.) of economic and cultural importance. This places a serious limitation on the ability of the archaeologist to infer the function of the site, season of occupation, and economic orientation, as well as other characteristics.

Thirdly, there is little doubt that surface collections are affected by "visibility biases." In other words, the representation of diagnostic archaeological materials found on the surface may be affected by landscape changes through time — most notably, the deep burial of early-to-middle Holocene sites by mass wasting processes. In the Sheyenne Delta of North Dakota, for example, Running (1995) argues that greater landscape instability (i.e., alluvial fan formation and eolian activity) characterized the mid-Holocene between 8,000 and 5,000 BP. Therefore, one would expect an inherent bias towards the over-representation of surface components that postdate the major period of instability.

The Geological Record

A previous study (Sun 1996) established the general history of glacial Lake Hind during the late Pleistocene. This work demonstrates quite clearly that, while glacial Lake Hind was relatively small (c. 4000 km²) in comparison to adjacent glacial Lake Agassiz, it was a part of the larger proglacial lake-spillway system. Lake Hind received meltwater from western Manitoba, Saskatchewan, and North Dakota via at least ten channels, and discharged into glacial Lake Agassiz through the Pembina spillway (Sun 1996). Therefore, the major geological events occurring within the Hind basin reflect a larger sequence of catastrophic proglacial lake drainage across the northern prairies (Sun 1996; Sun and Teller 1997). For this reason, the chronostratigraphic record preserved within the Hind basin has significance at the local and at the broader, regional scale.

For archaeological purposes, however, the work of Sun (1996) is of limited use. This largely stems from the absence of radiometric dates linking this geological record to an absolute timescale. Much of the literature on the North American proglacial lake-spillway system lacks reliable absolute dates (Elson 1983: 37). Indeed, many radiocarbon samples collected in the past consisted of limnic sediments shown to be contaminated, in some cases, by pre-Quaternary palynomorphs and weathered lignite (Nambudiri et al. 1980; Elson 1983: 37). To some extent, this

Table 1. Radiocarbon dates referred to in text. Depths refer to Figures 3 and 5.

Provenience; Depth	Lab. Number	Material/ Fraction	Radiocarbon Age (^{14}C yrs. BP)	2-δ Calibrated Range* (cal BP)
Upper eolian sand sheet paleosol (565 cm)	Beta-109529	Archaeological (hearth) organics	3250 ± 70 BP	3650–3350 cal BP
Mud above evaporite (485 cm)	Beta –109900	Bison long bone fragment	4090 ± 70 BP	4840–4440 cal BP
Lower eolian sandsheet (245 cm)	Beta-111142	Wood (*Salix*)	6700 ± 70 BP	7667–7462 cal BP
Upper gyttja (145 cm)	TO-7692	*Menyanthes* seeds	9250 ± 90 BP	10596–10228 cal BP
Lower gyttja (125 cm)	Beta-116994	*Menyanthes* seeds	10420 ± 70 BP	12677–11911 cal BP

*Calibration assessed using Stuiver and Reimer (1993). This calibration converts the radiocarbon age to "real" (i.e. calendar) years. The calibrated range contains all probable dates within two standard deviations (2-δ) from the mean.

problem has been alleviated by using only dates on wood and other plant macro-fossils (Clayton and Moran 1982). Despite this cautionary measure, however, the potential for error due to the reworking of older wood into more recent contexts remains a significant problem.

Additionally, the literature on the geology of the Hind basin is of limited archaeological use because it is almost entirely confined to the late Pleistocene (18,000–10,000 BP). Since the vast majority of the North American archaeological record is restricted to the Holocene (10,000 BP to present), only minimal overlap is available for integrating human history with landscape evolution in this region.

For the study area, a series of cutbanks located adjacent to the Souris River in the Lauder Sandhills promise a means of connecting the early record of human settlement to post-glacial landscape evolution in the southern Hind basin. These cutbank sites are unique because they expose a near-complete and widespread (c. 2 km minimum) sedimentological sequence spanning the terminal late Pleistocene to the present. This sequence is defined by the Flintstone Hill (FSH) site (Figure 1), which has yielded the major lithostratigraphic database used in this study. Rather than present all results generated from this site, a summary of the major paleoenvironmental changes for the periods of interest suffices for the purposes of this article. The full database is available in Boyd (2000). All relevant radiocarbon dates from the Flintstone Hill site are presented in Table 1.

A Brief Overview of the Folsom and McKean Archaeological Complexes

Folsom Complex (c. 11,000–10,000 BP)

The Folsom complex is recognized by its distinctive full-fluted spear points, although an unfluted variant — "Midland" — is also known (Wendorf et al. 1955; Hofman et al. 1990; Hofman and Graham 1998: 101). At excavated sites elsewhere in North America, these artifacts have been associated with a diverse tool kit that included small end scrapers, drills, biface knives, choppers, stone beads, as well as a range of bone tools (Dyck 1983: 75). The Folsom economy on the southern plains appears to be largely focused on bison hunting, although at the Owl Cave site (Idaho), mammoths may also have been procured in the beginning (Miller and Dort 1978: 129–39). At later sites, Folsom game also included mountain sheep, deer, marmots, rabbits and wolves as secondary components to a bison-dominated diet (Dyck 1983: 74).

Folsom surface recoveries are present across the Canadian Prairies only in small numbers. Vickers (1986: 35) suggests that the limited penetration of the complex may have been due to its short duration and "southern" (i.e., United States High Plains) origin. Pettipas and Buchner (1983: 421), on the other hand, argue that the scarcity of fluted Paleoindian points in Manitoba may be due to the failure of ice-marginal communities to attract significant populations of game.

McKean Complex (c. 4,000–3,000 BP)

The McKean complex subsumes three projectile point types (McKean, Duncan and Hanna) which were originally considered to be independent. Not long after their initial description, however, all three were lumped into a single complex because they were found mixed together at the type site in Wyoming (Dyck 1983: 100). The geographic distribution of the complex is quite extensive, with sites known throughout the northern prairies, aspen parklands, and the southern edge of the boreal forest (Dyck 1983). The number of McKean sites across the northern prairies is also quite large, in comparison with early Precontact materials such as the Folsom complex. In Alberta, for example, at least 139 McKean sites or isolated projectile points are known; Folsom, on the other hand, is represented by only three isolated points (Vickers 1986: figure 3).

The Cactus Flower site gives one of the best images of the McKean complex on the Canadian Prairies. This extensive site is located on the South Saskatchewan River north of Medicine Hat, in southeastern Alberta. The occupations associated with McKean, Duncan, and Hanna points yielded radiocarbon dates ranging from 4,130 ± 85 BP to 3,620 ± 95 BP. Other artifacts attributed to this complex include bifaces, gravers, endscrapers, spokeshaves, marginally retouched flake tools, pebble cores, hammerstones, anvils, a ground stone disk, a tubular pipe, bone awls, bone and shell beads, a shell disk, and an ammonite septum (Brumley 1975). The faunal assemblage from the site was dominated by bison (which indicated October/November and spring occupations from mandible age estimates), with smaller numbers of antelope, mule deer, canid, and trace quantities of kit fox, rabbit, birds, freshwater clam and fish (Brumley 1975). Based on a comparison of element numbers per individual, Brumley (1975: 83) suggests that bison were hunted nearby and antelope were taken from a greater distance. Repeated occupation of the Cactus Flower site by McKean groups was probably a result of the site's ideal location for opportunistic bison ambushing at a major river crossing (Brumley 1975: 91).

Late Pleistocene-Early Holocene Transitions in the South-Central Hind Basin: Lake Recession and Folsom Incursions

The Geological and Paleoecological Records from the Flintstone Hill Site

At the Flintstone Hill site, two basal units record the sequence of events occurring within the south-central Hind basin during the terminal late Pleistocene and early Holocene (Figure 3). The earliest of these, the glaciolacustrine unit, is approximately 1.2 m thick (above the level of the Souris River) and is composed of gleyed, carbonate-rich, massive to planar bedded clay to silty clay which grades upwards into an organic rich and finely laminated buried organic deposit (i.e., the "gyttja unit"). The c. 30 cm thick gyttja unit is composed of alternating planar beds of fine-texture clastics (clays and silts) and thin (c. 1–5 mm thick) detrital organic beds. Some occasional symmetrical ripple structures (≤ 5 mm high) are preserved in the organic bedding, suggesting periodic, low energy/low velocity flooding.

Together, the glaciolacustrine and gyttja units at the Flintstone Hill site are interpreted as showing a gradual transition from a deep-water to a shallow-water depositional environment (Boyd 2000). An accelerator mass spectrometry (AMS) radiocarbon date on seeds of the boreal emergent *Menyanthes trifoliata* (buckbean) recovered from the bottom of the gyttja indicates that this process was completed in the Flintstone Hill region by at least c. 10,400 BP (Figure 3). When linked to the model of glacial Lake Hind produced by Sun (1996), this sedimentological transition very likely corresponds to the drainage of the southern Hind basin following the catastrophic routing of meltwater through the Souris-Pembina spillway system during Phase 8. As illustrated in Figure 1, Sun's (1996) shoreline positions for Phases 7 and 9 indicate that the Flintstone Hill site was within the region of the Hind basin that was drained by this event.

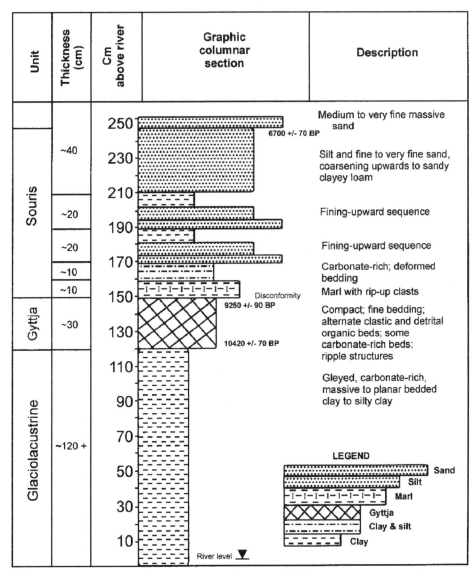

Figure 3. Composite Schematic Column of the Flintstone Hill Cutbank (Glaciolacustrine, Gyttja, and Souris Units), Southwestern Manitoba (Source: M.J. Boyd and G.L. Running IV, July 1998).

Plant macrofossils and microfossils preserved in the gyttja provide a sequence of paleoenvironmental change for the period between c. 10,400 and c. 9,300 BP (Boyd 2000). Based on pollen evidence, upland communities surrounding the Hind basin were initially dominated by *Picea glauca* (white spruce), *Artemisia* (sage), grasses, *Shepherdia canadensis* (Canada buffaloberry), *Populus* (poplar), and *Juniperus* (juniper), among others. At the same time, within the Hind basin in the area of the Flintstone Hill site, at least one small (< 1 km diameter), shallow, tree-less wetland remained following drainage. This small wetland was rapidly colonized by the aquatic and semi-aquatic taxa *Menyanthes trifoliata* (buckbean), *Carex* spp. and *Eleocharis* sp. (sedges), and *Potamogeton* (pondweed). Beginning shortly after c.

10,400 BP, glacial retreat and a warming climate are evidenced by the rapid deterioration of white spruce populations on the surrounding uplands. This process was briefly interrupted by a sharp cooling trend and a corresponding increase in the relative frequency of white spruce pollen. This subsequent Picea peak is seen in pollen spectra across southwestern Manitoba (e.g., Ritchie and Lichti-Federovich 1968; Ritchie 1969; Boyd 2000), and probably corresponds to a re-advance of the Laurentide ice sheet at c. 10,000 BP during the Emerson Phase of glacial Lake Agassiz (Teller et al. 1996: 62). By 9,300 BP, however, white spruce populations were virtually eliminated from the local uplands, leaving mostly *Populus*, *Juniperus*, and grasses. Within the Hind basin, at the Flintstone Hill sample location, the period following 10,400 BP was characterized by the dominance and subsequent decline (in terms of total seed numbers) of *Menyanthes trifoliata*. This emergent species probably dominated the local wetland due to highly fluctuating water levels produced by periodic, low energy, flooding (Haraguchi 1991; Boyd 2000). Presently, the distribution of *M. trifoliata* is mainly restricted to bogs and swamps in the boreal forest (Looman and Best 1987: 592).

The early postglacial forest reconstructed for the uplands surrounding the Hind basin was part of a larger, spruce-dominated assemblage (lacking a modern analogue) that spread across much of Canada's western interior immediately following deglaciation (Ritchie 1976: 1793). *Picea glauca* forests were established in North Dakota by c. 12,000 BP (Grimm 1995), and extended into southern Saskatchewan by at least 10,200 BP (Yansa and Basinger 1999: 151). Further west, open grasslands have apparently existed continuously in Montana over the past 12,000 years, with spruce being notably absent (Barnosky 1989). As well, in southeastern Alberta, *Populus* was present but *Picea* was probably absent (Beaudoin 1992). In North Dakota, the spruce forests were replaced by grassland vegetation at c. 10,000 BP (Grimm 1995) while, in southern Saskatchewan spruce forests were succeeded by deciduous parklands after c. 10,200 BP (Yansa and Basinger 1999: 151).

Folsom Site Distributions and Implications

The distribution of fluted Paleoindian (i.e., Clovis and Folsom) projectile points in southern Manitoba is largely restricted to the uplands located on the western side of the province (Figure 4). This pattern is sensible, given the fact that the eastern half was covered by glacial Lake Agassiz at this time (see Buchner and Pettipas 1990). In the study area, one Folsom projectile point from a private collection has a provenance within the Melita locality (Pettipas 1967: 356), and an unfluted ("Midland") form was recovered *in situ* from a river terrace outside of Souris, at an elevation of 442 metres above sea level (Pettipas 1967: 355) (Figures 1 and 4). These recoveries are significant because they are located well below the maximum outline of the glacial Lake Hind basin (i.e., 457 metres above sea level). More precisely, the Melita point was found in an area of the basin covered by water during Phase 4 (Figure 1). Although no radiocarbon dates are available for this early phase of glacial Lake Hind, it must postdate the initiation of deglaciation (i.e., after c. 12,000 BP [Sun, 1996]). The Souris specimen, furthermore, was found in an area of the basin that would have been covered by water during, and prior to, Phase 7 (Figure 1). Since the Phase 7 to 9 transition is known to have been completed shortly before c. 10,400 BP, the drainage of this region was probably either well underway, or entirely completed by Folsom (11,000–10,000 BP) times.

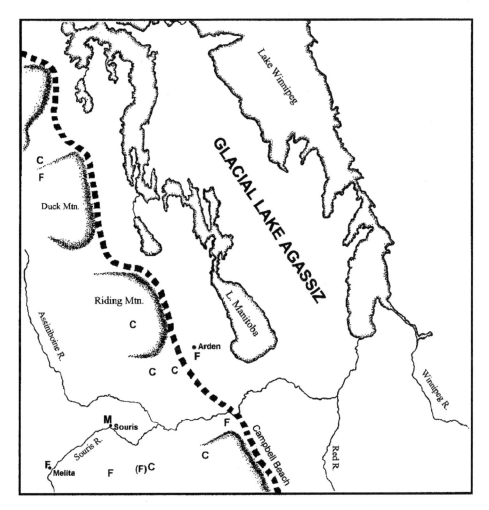

Figure 4. Early Paleoindian Surface Recoveries in Southern Manitoba (C = Clovis, F = Folsom, M = Midland, (F) = authenticity disputed).

Significantly, on the western shore of glacial Lake Agassiz, the purported recovery of a Folsom point near Arden — below the elevation of the Campbell beach (Figure 4) — is consistent with a contemporaneous drop in the level of Agassiz during the low water-level Moorhead Phase (c. 11,000–10,000 BP) (Buchner and Pettipas 1990: 53).

From this evidence, therefore, it seems that at least one component of the Folsom land-use strategy in southwestern Manitoba included the utilization of recently drained proglacial lake surfaces. Since the paleoecological data from the Flintstone Hill site indicate that at least one residual wetland in the Hind basin was a low-diversity, treeless fen — assuming the sample location is broadly representative of similar contexts — it seems unlikely that these drained regions would have provided an adequate hunter-gatherer resource base in themselves. Instead, given indications at several Plano[2] Paleoindian sites of a bison mass-drive and entrapment hunting technique in association with natural barriers (e.g., sand dunes [Frison

1971], arroyos [Wheat 1972], ponds [Sellards et al. 1947], and glaciolacustrine sands at the Fletcher site in Alberta [Forbis 1968; Quigg 1976; Vickers and Beaudoin 1989]), these areas may have provided reliable opportunities for the planned entrapment of bison in wet clay beds and spillway channels. Although a Folsom bison kill site in Manitoba has yet to be found, this argument is supported in a general way by the location of early fossil bison remains in the province.

The presence of fossil bison in association with the proglacial lake-spillway system is reasonably well established. In the Swan River valley of west-central Manitoba, for example, fossil bison remains have been recovered from gravels on an extensive spit complex situated between the Upper and Lower Campbell strandlines of glacial Lake Agassiz (Nielsen et al. 1984). These remains are, morphologically, rather large and were assigned to either *Bison bison occidentalis* or *B. bison antiquus* by the original analysts (Nielsen et al., 1984: 834). Three radiocarbon dates were obtained from these materials: 10,300 ± 200, 9400 ± 125, and 9,500 ± 150 BP (Nielsen et al. 1984: 832). Other bison remains of the "*occidentalis*" phenotype have also been recovered from middle Holocene fluvial contexts in southern Manitoba (e.g., Dyck et al. 1965; Steinbring 1970; Nielsen et al. 1996: 12–13). These recoveries suggest that the former range of early bison, at least in part, included lake margins, rivers, and other wetland locales. This pattern, furthermore, was probably established by at least Folsom Paleoindian times (Nielsen et al. 1984: 838). On this basis, the Folsom materials recovered from within the glacial Lake Hind basin may indicate a deliberate exploitation of this behaviour. Given the paucity of Folsom projectile points in southern Manitoba, this strategy may have been a relatively late and short-term development, perhaps coinciding with the initial deterioration of *Picea glauca* woodlands prior to the Emerson Phase glacial readvance (i.e., shortly before c. 10,000 BP).

The McKean Complex and the Development of Prairie in the Hind Basin

The Middle Holocene Geological and Paleoecological Records

Following c. 9,300 BP, but prior to c. 6,700 BP, a series of crude fining upward sequences of clays, silts and some sands was deposited in the area of the Flintstone Hill site (Figure 3). Based on the radiocarbon constraints, as well as the relative sequence of events outlined in Sun (1996) and Sun and Teller (1997), these sediments are probably alluvium, deposited after the initial incision of the Souris River into lacustrine deposits (Boyd 2000). After c. 6,700 BP (c. 7,600 cal BP[3]), at least one eolian sand sheet was deposited across the 2 km study area. The lowest of these overlies the Souris unit (i.e., the alluvium) without evidence of an erosional unconformity. Instead, a coarsening upward sequence is indicated: from gleyed silts and clays, to silty sand, to the sand-dominated lower sand sheet subunit (Figure 5). At the contact between the Souris unit and the lower eolian sand sheet subunit, plant macrofossils (which yielded a date of 6,700 ± 70 BP [Table 1]) are abundant, and consist mostly of leaves from the thermophilous tree species *Quercus macrocarpa* (bur oak), and seeds of *Cornus stolonifera* (red-osier dogwood), *Vitis riparia* (riverbank grape), *Sparganium* sp. (bur-reed), and *Rubus idaeus* (raspberry) (Boyd 2000). The pollen profile, while indicating the predominance of *Quercus* (35–85 percent), also attests to the importance of *Populus* (5–40 percent) and grasses (2–15 percent), and the local presence of *Salix* (willow) (0–10 percent) (Boyd 2000). Based

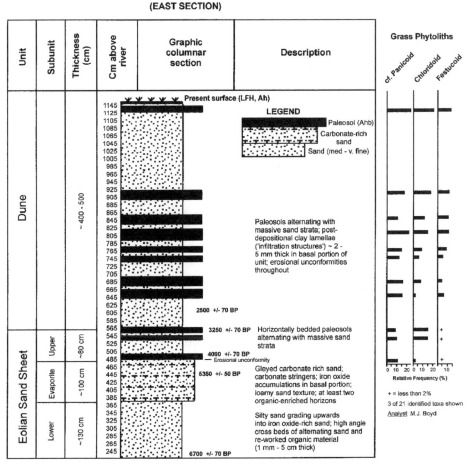

Figure 5. Composite Schematic Column of the Flintstone Hill Cutbank (Eolian Sand Sheet and Dune Units), Southwestern Manitoba (Source: M.J. Boyd and G.L. Running IV, July 1998).

on the range of moisture regimes implied by the plant assemblage, a mosaic of dry (i.e., prairie), mesic/riparian oak forest, to wetland habitats probably lined the inside edges of the south-central Hind basin (in the vicinity of the Souris River) by c. 6,700 BP. While no paleosols (buried soils) have been found in the lower eolian sand sheet subunit, some alternating, thin (1 mm–5 cm), high-angle cross beds of sand and reworked organic duff were observed (Boyd 2000).

Following c. 6,700 BP, the development of the Oak Lake aquifer in the south-central Hind basin is strongly indicated in the lower eolian sand sheet and evaporite subunits (Figure 5). These subunits indicate a progressively rising water table for at least part of the middle Holocene on the basis of the following: 1) post-depositional, upward-contorted, accumulations of iron oxide in the lower eolian sand sheet subunit; 2) the presence of thick gleyed sediment in the overlying evaporite subunit; and 3) the appearance of iron oxide accumulations below the concentration of more soluble carbonate minerals in the evaporite subunit, implying upward water movement (Figure 5) (Boyd 2000). The rather profound accumulation of carbonate minerals in the evaporite subunit suggests mineral concentration by evapotranspiration of groundwater (Boyd 2000).

Together, the lower eolian sand sheet and evaporite subunits suggest that the period following c. 6,700 BP was characterized by an initial increase in effective precipitation (producing a rising water table), followed by a period of drought (producing high rates of evapotranspiration and the evaporite deposit). Significantly, this trend mirrors recent evidence from Ingebrigt lake (southwestern Saskatchewan) for high relative humidity between 7,200–6,500 BP, followed by a period of much lower relative humidity between 6,500–5,500 BP (Shang and Last 1999: 107).

Following 4,090 ± 70 BP (c. 4600 cal BP), but prior to 3,250 ± 70 BP (c. 3,500 cal BP), a series of significant changes were initiated in the study area (Figure 5). By this time, at least two parallel planar-bedded paleosols formed across the site, interbedded with massive sand strata (Figure 5). These paleosols (c. 10–20 cm thick) consist of simple, black, Ah-horizons, suggesting development under predominantly grassland vegetation (Dormaar and Lutwick 1966; Buol et al. 1989). This pedologic interpretation is supported by a grass-dominated microfossil assemblage associated with the buried soils, and a simultaneous absence of arboreal indicators. Grasses, in this analysis, were identified on the basis of diagnostic phytolith forms produced by the precipitation of hydrated silicon dioxide inside the living plant tissue (see Brown 1984; Twiss 1992). As Figure 5 shows, grass phytoliths from the upper eolian sand sheet paleosols are mostly derived from the Chloridoideae (warm/dry short grasses) and Panicoideae (warm/moist tall grasses) subfamilies. This is consistent with the modern composition of grasses in the open prairie and sandhill locales across the study area; in contrast, as summarized above, the regions covered by open *Populus tremuloides* forest tend to have a greater Festucoideae native grass component.

These lines of evidence suggest that the period between c. 4,100 and 3,300 BP was characterized by extensive eolian landscape stability and the widespread growth of prairie taxa on uplands in the south-central Hind basin. The trend towards increased landscape stability during this period may be due to the onset of higher relative humidity between c. 5,000 and 2,000 BP on the northern prairies (Shang and Last 1999). This rather lengthy period of landscape stability and relatively moist conditions is recorded in many basins across the region (see summaries in Lemmen 1996; Vance 1997; Xia et al. 1997).

In contrast, between c. 6,700 and 4,100 BP, the Flintstone Hill lithostratigraphic record shows little evidence of landscape stability. Instead, this period was characterized by the deposition of at least one eolian sand sheet, a rising water table and the development of a playa system, followed by high rates of groundwater evapotranspiration. These conditions may have prohibited, in the long run at least, widespread prairie growth across the south-central Hind basin. Instead, for at least part of the middle Holocene, grassland vegetation may have been far more "patchy" across the study area due to circumscription by extensive playas and greater eolian activity. This model would imply a far more lengthy process of grassland succession for the Hind basin as a whole, in contrast with upland regions on the northern plains. In North Dakota, for example, open prairie rapidly replaced the early postglacial spruce forests at c. 10,000 BP (Grimm 1995). In southern Saskatchewan and Manitoba, spruce forests were succeeded by deciduous parklands prior to establishment of open prairie at c. 8,800–9,000 BP (Ritchie and Lichti-Federovich 1968; Yansa and Basinger 1999).

McKean Complex (c. 4,000–3,000 BP) Site Distributions and Implications

McKean sites are scattered across much of southern Manitoba, although some prominent clusters are apparent (Figure 6). These concentrations seem to be broadly associated with modern lowlands, rivers, wetlands, and lakes (e.g., Swan River Valley [I], the Hind basin [II], Pembina River and Rock Lake [III], and the Winnipeg River system [IV]). Although, in small part, these site distributions may reflect some collection biases (e.g., slightly higher concentrations near towns and cities, differences in site visibility, etc.), the apparent paucity of sites on higher areas within the Manitoba Escarpment suggests that other factors are involved. This must be the case since the Glacial Lake Agassiz Survey — the major effort responsible for mapping the majority of McKean surface sites shown in Figure 6 — also examined these more elevated regions. For this reason, the distribution of McKean complex sites in Manitoba is mostly accepted as an artifact of real settlement patterning, rather than simply a collection bias (Syms 1970).

Figure 6. Location of McKean sites in southern Manitoba.

In the context of the geological and paleoecological reconstructions presented above, the duration of the McKean complex is clearly contemporaneous with the initiation of widespread landscape stability and the more extensive growth of prairie vegetation in the study area. Since there is also strong evidence for the development of a high water table (i.e., the Oak Lake aquifer) prior to the late Holocene, by at least 4,000 BP the region had probably developed into a large mosaic of wetland and prairie. Since the Hind basin seems to be associated with a regional cluster of McKean sites, deliberate selection for this type of landscape may be indicated. Indeed, close proximity to an ancient marsh or slough is found at both the Cherry Point (Haug 1976; #5 in Figure 6) and Vera sites (Nicholson and Hamilton 1998; #4 in Figure 6). Significantly, Wright (1995: 316) observes that McKean bison kill sites across the Canadian Prairies as a whole tend to be associated with modern sloughs (Wormington and Forbis 1965; Haug, 1976), river valleys (Adams 1976), tributaries (Wettlaufer and Mayer-Oakes 1960), and major moraines (Kelly and Connell 1978). The choice of site location probably reflects several criteria. To some extent, as suggested for the Cactus Flower site (Brumley 1975), these locations may have been key points on the landscape that afforded reliable opportunities for the entrapment of bison along their seasonal migration routes.

Conclusions

Southwestern Manitoba was a very different place from Folsom to McKean times. In the study area, Folsom hunter-gatherers would have encountered a proglacial lake which, by c. 10,400 BP was confined to only the deeper, north-central portion of the Hind basin (i.e., the Phase 9 boundary [Figure 1]). At some point following the gradual recession of glacial Lake Hind, Paleoindian populations probably made limited incursions into recently drained areas within the basin. These locales may have provided reliable opportunities for the entrapment of bison in wet clay beds, spillway channels, and other landscape obstacles. To some extent, this hypothesis is supported by the recovery of early fossil bison remains on Holocene floodplains (Dyck et al. 1965; Steinbring 1970), and adjacent to the western margin of glacial Lake Agassiz (Nielsen et al. 1984). Radiocarbon dates indicate that bison were exploiting these locales in Manitoba by at least 10,100 BP (Nielsen et al. 1984: 838).

The Folsom land-use strategy suggested in this study may have been part of a larger pattern. In west Texas, for example, Boldurian (1991: 285) argues that *Bison bison antiquus* probably established patterns of movement along deeply incised fluvial systems, "where an adequate supply of water and vegetation existed." Assuming that Folsom groups in the southern plains focused mainly on bison procurement, their seasonal rounds would have coordinated with major movements of these herds (Broilo 1971; Boldurian 1991: 285). It is perhaps not coincidental that many of the major river valleys in west Texas also pass through the source areas for the predominant siliceous stone materials used by Folsom hunter-gatherers in this region: Edwards chert, Alibates agate, and Tecovas jasper (Boldurian 1991: 284–85). As in most regions, however, Folsom settlement pattern studies in Texas are hampered by statistically insignificant projectile point distribution data (Largent et al. 1991).

In southern Ontario, Paleoindian settlement strategies have been linked to strandlines and beaches of former glacial lakes (e.g., Storck 1982; Stewart 1984;

Ellis and Deller 1990). An association with the contemporaneous strandline of glacial Lake Algonquin appears particularly strong (Storck 1982; Deller et al. 1986); notably, approximately 66 percent of sites along this physiographic feature occur near "complex" (i.e., indented rather than straight) strandline features such as former lagoons, islands, peninsulas, and embayments (Storck 1982: 19). Such locales may have provided access to larger areas of shoreline habitat within a short radius of the site than would locations on straight coastlines (Storck 1982: 23). This strong "orientation" to the strandline, moreover, is also supported by the fact that all of the sites have unrestricted visibility of this feature and, for approximately 40 percent of the sites, visibility is restricted by the surrounding terrain to the strandline only (Storck 1982: 23).

By the end of the middle Holocene (c. 4,100–3,300 BP), a period of greater landscape stability was initiated in the south-central Hind basin. This period was characterized by the development of more extensive grasslands above the Oak Lake aquifer, creating what may have been a very large mosaic of wetland and dry prairie. Since the Hind basin appears to be associated with a regional cluster of McKean sites, deliberate selection for this prairie wetland-riverine locale may be indicated. This, in turn, argues that the larger spatial association between McKean sites and modern lakes, sloughs, and river valleys across the Canadian Prairies may indeed be a faithful reflection of the original land-use pattern. Given the evidence of repeated bison kills at several McKean sites in Canada (e.g., Brumley 1975; Haug 1976), these locations may have been key points on the landscape that afforded reliable opportunities for the entrapment of bison along major migration routes. It is also important to note, however, that these locales may have been attractive for a number of additional reasons: e.g., aquatic/semi-aquatic plants and animals, and drinking water in the summer; firewood and windbreaks during the winter, etc. More than any one reason, it is perhaps this characteristic — i.e., the intersection of a number of different types of resources — which made these locales important points on the Precontact "economic landscape."

Despite significant changes in landscape and vegetation from the terminal late Pleistocene to the end of the middle Holocene in the Hind basin, it is interesting to note that Folsom and McKean hunter-gatherers may have spatially "mapped on" to the physical world according to a highly similar, and successful, economic strategy. At least in part, this shared strategy may have involved the entrapment of small numbers of bison in association with wetlands and river/spillway systems. Thus, while the precise location of sites may not necessarily be the same from Folsom to McKean times, a common land-use strategy may still have existed.

Acknowledgments

This research was funded through doctoral scholarships awarded by the Natural Sciences and Engineering Research Council of Canada (NSERC) and Killam Trusts of Canada. In the field, support was obtained through a SSHRC grant (#410–97–0180) to B.A. Nicholson (Brandon University), in addition to support from the Manitoba Heritage Grants Program (#97F-W111-R), and Brandon University Research Council (#2666). Garry Running (University of Wisconsin –Eau Claire) and his students are gratefully acknowledged for their generous help in the field; many additional thanks are also owed to Garry for helping with the interpretation of the geological record. Bev Nicholson and Scott Hamilton (Lakehead University) deserve warm recognition for their on-going collaboration, and for establishing and maintaining a work environment that supports the friendly integration of both geoscientific and archaeological research goals. This article has benefited from conversations with Brian Kooyman, Len Hills, and Gerry Oetelaar (University of Calgary), and Leo Pettipas

(Manitoba Museum of Man and Nature). Len Hills is also acknowledged for funding one of the radiocarbon dates (TO-7692) under his NSERC grant. Finally, I am grateful to Dave Sauchyn and two anonymous Prairie Forum reviewers for their helpful comments. All remaining errors and/or omissions are solely the responsibility of the author.

Notes

1. A complex is a consistently recurring assemblage of artifacts or traits which may be indicative of a specific set of activities, or a common cultural tradition.
2. The Plano complex, originally defined by Jennings (1955), is a general term used to refer to several, presumably distinct groups recognized by unfluted, lanceolate projectile points. These materials post-date the Clovis and Folsom types (i.e., c. 10,200–8,000 BP), and are found throughout the United States Great Plains and Canadian Prairies.
3. This value is the approximate mean of the calibrated range presented in Table 1.

CHAPTER 3

CHAPTER 3

SOILS, DUNES, AND PRAIRIE VEGETATION: LESSONS FROM THE SANDHILLS OF NORTH DAKOTA

D.G. Hopkins and *G.L. Running*

ABSTRACT. Native vegetation of the Northern Great Plains evolved over millennia in a harsh and highly variable climatic regime, under the influence of grazing by herbivores. Conversion to agricultural production in this region occurred in considerably less than a century, and was often driven by external economic factors that did not include physical and biotic limits of the landscape. The history of land use in Sandhills of southeastern North Dakota illustrates the severe socioeconomic repercussions that occur when agricultural development is imposed on landscapes dominated by sands and frequented by drought. This article examines the soil paleoenvironmental record of the Sandhills in the context of historic climatic fluctuations that have affected the region. Stratigraphic evidence, soil properties, and radiometric dates of buried soils (paleosols) are presented that document cycles of erosion and recurring intervals of soil formation to at least 3,000 years before present. The resiliency of the native prairie vegetation to adapt to changes in groundwater elevation is shown by groundwater monitoring and vegetation analyses taken through the last decade. The intrinsic diversity of the native vegetation and relatively rapid rates of primary succession on bare dunes may explain the presence of multiple paleosols discovered at this site.

Introduction

Most soils in central and eastern North Dakota are naturally fertile due to their relatively young glacial parent material (Brady 1990) and a sub-humid continental climate that fosters the dominance of mixed-grass and tallgrass prairie vegetation (Barker and Whitman 1989). This vegetation community evolved under grazing on the prairies (McNaughton 1974) and is adapted to a harsh and highly variable climatic regime (Van Dyne 1979). The "prairie profile" resulting from this combination of soil-forming factors has high levels of organic matter, a sufficiency of divalent cations, and stable soil structure. Not all soils in the prairie regions, however, exhibit these "prairie profile" characteristics or are adaptable to crop growth. Some of the quarter million people who entered North Dakota between 1898 and 1915 learned this lesson the hard way (Robinson 1966), particularly on landscapes dominated by sands. These soils are quite susceptible to soil erosion because they lack adequate water holding capacity and the resilient soil structure that characterize finer-textured soils. Such sandy landscapes on the Northern Great Plains (NGP) are at best fragile when cultivated.

Historical periods of economic development on the North Dakota prairies, and across the NGP, resulted from a complex combination of factors. Prominent among these were federal programs such as the various Homestead Acts in the United States and the infrastructure development that followed the growth of railroads (Robinson 1966). Expansion in the early part of this century occurred during an opportune interval in a cyclic climatic regime not recognized by our predecessors. Robinson (1966, p. 236) stated that "the country was filled with people who did not understand it very well"; some would become unwilling victims of what historians refer to as the "boom to bust" syndrome on the NGP (Danbom 1990). External economic factors associated with this syndrome have been well documented (Robinson 1966), and a clearer understanding of the highly variable and drought-prone NGP climate now exists (Katz and Brown 1992). Such changes in climate were responsible for profound changes to both the physical environment and agriculture-based communities on the NGP; sandy landscapes under cultivation were markedly affected.

Unfortunately, strategies to anticipate the spatial and temporal variability of the NGP have been rarely implemented. All too often NGP policy makers and producers have ignored the "environmental imperative" or developed strategies to overcome, rather than adjust to, environmental constraints (Lemmen et al. 1998). These approaches are often expensive, destined to fail, and contribute to the "boom to bust" syndrome so characteristic of the region. We present a case study from the sand-dune dominated landscape of the Sheyenne Delta (the Sandhills), located in southeastern North Dakota, to illustrate our point. The physical environment of the Sandhills resembles many dune-dominated landscapes on the NGP, and land use trends in the Sandhills are also characteristic of post-settlement human-environment interactions across the NGP. The purpose of this paper is to: 1) use the detailed paleoenvironmental record preserved in the Sandhills to characterize the local soil-geomorphic system, 2) examine the relationship between climatic variability and landscape/vegetation response, and 3) identify the ecological and land-use implications of these relationships.

Environmental Setting

Landforms and Landscape

The largest contiguous area of sandy soils in North Dakota, known locally as the "Sandhills," exists on the delta formed when the ancestral Sheyenne River flowed into the southern Lake Agassiz basin. The Sheyenne Delta extends across 2,000 km^2 of Richland, Ransom, Sargent, and Cass counties. It is the second largest deltaic landform of Glacial Lake Agassiz in North America, exceeded only by the Assiniboine Delta of Manitoba (Brophy and Bluemle 1983). The ecological history and agricultural potential of the Sheyenne Delta is intrinsically linked to the sediments deposited by the Sheyenne River during late-Wisconsin time and their subsequent reworking by geomorphic processes since deglaciation. The delta resembles a sediment wedge 12 metres thick near the western origin and 55 meters thick at the delta's eastern edge (Baker 1967). Sediment size varies in roughly concentric bands on the delta, from very coarse sand on the west to very fine sand and silt on the eastern edge (Figure 1).

A variety of landforms composed of sandy sediments reworked from underlying deltaic deposits dominate the delta surface (Harris 1987; Running 1997). All of the eolian landforms on the delta surface today are consistent with the parabolic dune

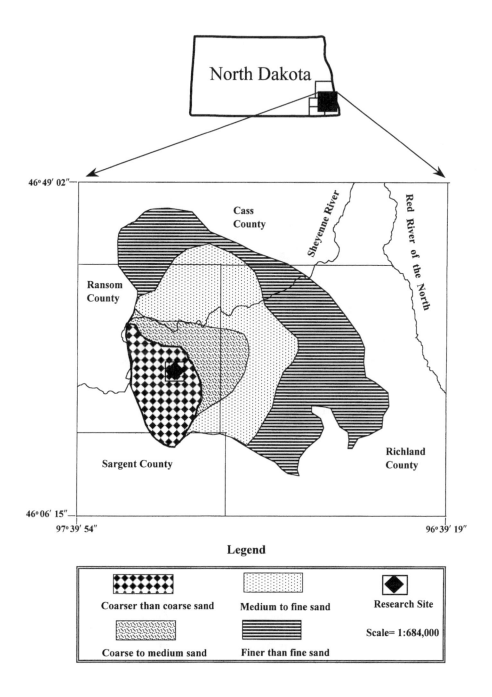

Figure 1. Location of the Sheyenne Delta in North Dakota and texture of deltaic sediments (adapted from Baker 1967). The research site is in the western part of the Sheyenne Delta.

association as described by David (1979) for the Canadian prairies. Locally, topographic expression on the delta surface reflects the underlying dune landforms present.

The western half of the delta is characterized by hummocky topography consisting of closely spaced, closed or irregular depressions and intervening knolls and low mounds. To most people living in the area, and for the purposes of this discussion, the "Sandhills" is synonymous with this part of the delta. Most of this part of the delta has never been cultivated (Burgess 1965) making it "the most extensive remnant of tallgrass prairie" in eastern North Dakota (Wanek 1964). The hummocky topography reflects the presence of a variety of low-relief parabolic dune types that are largely stabilized. These dunes range from crescentic to linear in form, 1-5 meters in height, and from 50 to 250 meters long. They are most extensive surrounding or downwind of flat, irregular-shape closed depressions referred to as deflation areas. Deflation areas supplied the sand for dune formation. These dunes exhibit irregular surfaces resulting from sand blowouts, especially along crests and ridges. Evidence suggests they formed under climatic and vegetation conditions similar to the present (Running 1996; Running and Boutton 1996) and have repeatedly experienced short periods of instability. The dunes are exceptionally vulnerable to erosion when cultivated or overgrazed.

The eastern and southern portions of the delta surface have a subdued, gently undulating surface. This topography results from the presence of an eolian sandsheet that is 3 meters thick or less and composed of finer sands and silts than the hummocky landscapes to the west. The eastern part of the delta surface is extensively cultivated; and the arc of stable farming communities located there (Kindred, Wyndmere, Barney) affirms the higher soil productivity conferred by finer sediment size and gentler topography (Figure 1).

Soils and Groundwater

The earliest soil survey of the western Sandhills identified the Fargo and McLeod sands as the dominant soil mapping units (Ely et al. 1907). Sand content for surface soils was about 85 percent for the Fargo sand and over 90 percent for the McLeod sand. McKinstry (1910) mapped a small area near McLeod and measured sand contents over 90 percent for most of the soils. He specifically warned that the "upland" soils found on hillslope and crest positions of dunes had a "light sod" and that "blowouts" could occur where this sod was disturbed. Nelson (1986) performed the most recent analysis of soil texture in the vicinity of the current study, and found an average of 90 percent sand and slightly less than 10 percent silt plus clay for upland soils. In a broader Sandhills grassland survey, Wanek (1964, p. 9) noted that soil profile development "is primarily related to the amount of organic matter incorporated into the upper soil horizons." Thirty sites were chosen in "upland prairie," and triplicate soil samples analyzed for soil organic matter showed a mean of 1.3 percent. Water retention was determined in the field 2 days following heavy widespread rains. The mean field soil water content was only 8.6 percent, which Wanek attributed to the extremely high (94.6 percent) average sand content. Wanek identified four ecological stages in blowouts and stated that edaphic properties (organic matter, soil nutrients, and water retention) were most limiting in pioneer stages. Water status improved for the next two stages, and was highest in the climax stage.

The major source of groundwater in the Sandhills is the Sheyenne Delta aquifer. This extensive, relatively shallow aquifer (750 mi^2) is the largest in Ransom County. Ely's survey crew noted that water was "never very far from the surface" in the lowlands, and even late in the fall during normal drawdown, groundwater was less than 5 feet from the surface (Ely et al. 1907, p. 34). The major source of recharge to the aquifer is precipitation that percolates through the veneer of sandy soils, producing a calcium-bicarbonate water type with very low salinity. The slope of the water table in the aquifer in areas beyond the Sheyenne River is very low, and local gradients, "which are toward individual low areas where evapotranspiration is greatest, mask regional trends" (Armstrong 1982, p. 33).

Land Use in the Sandhills

Although the plow is responsible for much of the devastation, a large part of it is due to no practice other than severe and continuous overgrazing." (Anonymous)

The first soil survey of the western Sandhills emphasized both variable relief and depth to groundwater as crucial determinants of agricultural potential. The problem in farming McLeod sand, and much of the Sandhills proper, was perceptively attributed to "the peculiar topography", which "makes cultivation somewhat difficult, owing to the fact that while the top of the knoll may be very dry the depression may be very wet, and it is hard to find a season that suits both" (Ely et al. 1907, p. 18). The high sand concentrations determined by Ely et al. (1907) and McKinstry (1910) were corroborated by later researchers using better tools. Water holding capacity is nil in such sands, creating droughty conditions for upland soils, while relatively shallow groundwater insures a sufficiency (or excess) of water at lower landscape positions. Perhaps the most astute observations made by McKinstry (1910) concerned the linkage between the vigorous prairie vegetation in lower landscape settings and the presence of capillary water, which enhanced soil productivity. Root proliferation was recognized as much weaker on fine sands in upland positions, but little drifting was noted at the time of the soil survey "unless the sod had been destroyed" (Ely et al. 1907, p. 18). The merits of perennial grass roots for binding sand as well as the dangers of overgrazing were noted by Bell (1910) in his appraisal of the Biological Survey of the McLeod area. McKinstry (1910, p. 110) was unequivocal in his opinion regarding farming potential for the McLeod area:

> Most of this land is lying idle at the present time and is increasing in value very slowly, so there is practically nothing to be gained by holding it. In fact, so much money has been lost by investments in the last few years that to secure loans on any of the land is almost an impossibility.

In spite of the caution urged by those early resource specialists, population expanded in the McLeod area. Part of the reason for the initial growth was that average rainfall was higher in the later part of the 1800s and the early decades of the twentieth century (Figure 2), which provided higher than average yields on marginal lands. This trend for higher precipitation was consistent throughout southeastern North Dakota (Dr. John Enz, North Dakota State Climatologist, pers. comm. 2000) and much of the Canadian prairie provinces (Lemmen 1996). Producers may have been lulled into believing the landscape was more benign than they anticipated, a significant error on sandy soils. National and international economic priorities certainly played a role in development. For example, North

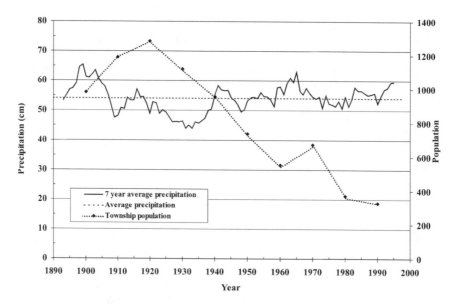

Figure 2. Seven-year running average of yearly precipitation at Wahpeton, ND compared to 1890-1998 yearly average and population in the 6 civil townships closest to McLeod, ND.

Dakota Governor Frazier urged farmers to plant every available acre the day after the declaration of war in 1914 (Robinson 1966). High prices for agricultural commodities during World War I prompted expansion of new farms and businesses in the Sandhills. Establishment of the Federal Land Bank facilitated the "boom" but farmers borrowed heavily during the 1920s and overgrazing was apparently widespread, as farmers were forced to maintain herds due to low cattle prices (Robinson 1966; Thorfinnson 1975). Overgrazing was known to reduce revegetation rates in blowouts (Shunk 1917), and the decline in precipitation during the late 1920s (Figure 2) additionally deteriorated rangeland condition.

The 1930s drought exacerbated economic effects of the Great Depression and both the natural and social fabric unraveled in the Sandhills. The McLeod locality was particularly devastated. Drifting sand covered former fields, shelterbelts were buried, and dozens of farms were delinquent in taxes (Thorfinnson 1975). Population in the 560 km² area around McLeod began its steady decrease to the present (Figure 2), a decline mirrored across many parts of the NGP.

President Roosevelt's Commission on the Future of the Great Plains called for "an intelligent readjustment to the ways of nature" following the dust bowl (Robinson 1966, p. 417). The federal government initiated a Land Use Project to ameliorate 34,600 ha of submarginal lands in the Sandhills; about 55 percent of this land was in arrears for taxes. Between 1937–39, the federal government purchased 27,244 ha under the Bankhead-Jones Farm Tenant Act (Manske et al. 1988). All the foreclosure land was later mapped as fine sands or coarser by Baker (1967). Rangeland stabilization programs were initiated by the Soil Conservation Service in 1940; and in 1941 the Sheyenne Valley Grazing Association was formed to promote sustainable range management practices. Administration of the federal land was transferred to the United States Forest Service (USFS) in 1954 (Wanek 1964; Manske et al. 1988); the Forest Service now manages the Sheyenne National

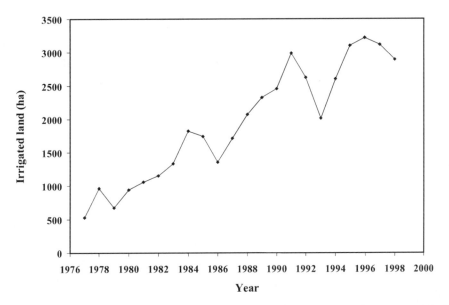

Figure 3. Increase in irrigated land in the Sandhills of southeastern North Dakota (source: ND State Water Commission).

Grasslands (SNG) under a multiple-use policy, but grazing by livestock is the dominant land use (Nelson 1986).

In the 1990s, economic opportunities due to the developing potato industry resulted in land use changes in the Sandhills, and in other sand plains regionally, that may increase the potential for soil losses due to wind erosion. Prairie vegetation provides the best protection against wind erosion on sandy soils, while cereal grains offer good protection. However row crops such as corn, sunflowers and potatoes expose soil to damaging erosion, especially in the windiest months of April and May. The Ransom County acreage planted to irrigated potatoes, including the Sheyenne Delta, was virtually nil in the mid-1980s, but increased to 1300 hectares by 1998 (North Dakota Agricultural Statistics Service 1999). Irrigation development rose markedly, and applications to withdraw water from the surficial aquifer continue to increase (Figure 3). If excessive groundwater drawdown occurs, the vigor of the native vegetation may be weakened, which could increase soil erosion on the prairie.

Lessons from the Land

In the past, some Sandhills producers have made land use decisions based on perceived economic gain without understanding the physical and biotic limits of the local ecosystem. Land use changes proceeded above the soil surface on a temporal scale of years and decades, while a hidden chronicle of environmental history spanning thousands of years lay encapsulated in the sandy profile.

Buried soils are commonly observed in dune exposures throughout the Sandhills, but this soil-stratigraphic evidence of repeated cycles of denudation and quiescence went unnoted in early soil and vegetation studies. It was Wanek (1964, p. 11) who mentioned that buried profiles were a "common occurrence" in the

Figure 4. Location of transect linking Soo Dune and Maddock Dune paleosols; Sandoun Township, Ransom County, ND. Paleosols shown by diamonds, instrument positions by triangles (base map compiled from Venlo, ND and McLeod, ND 7.5 Min. USGS topographic maps).

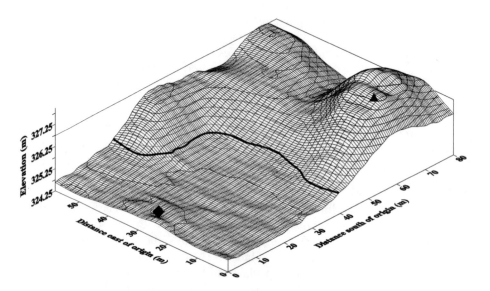

Figure 5. Landscape near Maddock Dune paleosol looking southeast. Paleosol location and origin of transect shown by filled triangle. Point IP1 on transect is ~ 10 m east of knoll at southeast corner of grid. Groundwater monitoring well shown by filled diamond. High water line due to ponding since 1993 is shown by bold contour line (vertical exaggeration x 7).

Sandhills and that "three and four profiles in a vertical sequence were observed in several instances." The frequent appearance of buried profiles on exposed sand dunes was also noted by Burgess (1965), and geologists stated that "multiple paleo-regosols occurred in the eolian sequence" (Brophy 1967, p. 162). Bluemle (1979) noted that some buried soils on the Sheyenne Delta have more developed profiles than modern soils. The ages and environmental implications of these buried soils were not investigated until recently (Hopkins 1997; Running 1996; Running 1997).

Buried soils (paleosols) can be important environmental indicators because they document intervals of landscape stability during which soil development occurred. Subsequent episodes of landscape instability can either truncate existing soils or cover them with new sediments; both processes essentially "reset the clock" for soil development. Paleosols indicate past climatic conditions and vegetation because: 1) the degree of profile development is roughly proportional to the length of the stable interval, and 2) the physical and chemical properties of soil horizons are strongly controlled by vegetation. Thus, buried soils (or even better, sequences of several buried soils), where present, can be a powerful tool for interpreting the past. Their age can be determined using carbon isotopes extracted from soil organic matter; they bracket periods of environmental change.

Two sites from the western Sandhills with paleosols were dated recently by independent research teams and results show disparate ages for the two sites, despite their proximity (Figure 4). These paleosols vary in thickness, depth, organic matter content, soil color, and in age. (Hopkins 1997; Running 1997 1997).

The Maddock Dune sequence includes a single paleosol, observed 2.05 m below the dune crest (Figure 5), which is relatively thick (0.33 m) compared to buried soils from nearby dune exposures. Chemical determinations for soil organic matter showed distinct increases for the Maddock paleosol compared to eolian sands above it (Table 1). This buried soil developed in light yellowish brown sands, but a weakly developed brown B horizon about 20 cm thick was observed below the 2Ab horizon (Table 1). The color of this subsurface horizon indicates that iron-bearing minerals were oxidized during soil formation, which requires a period of perhaps a few hundred years. The lack of drab olive hues below the 2Ab horizon suggests this soil was reasonably well drained; it probably did not develop on the lowest position of the paleo-surface. The entire soil profile was leached of calcium carbonate above the 2Ab horizon. Radiocarbon analysis of the Maddock Dune paleosol provides an age of 2370 +/- 60 BP (Hopkins 1997).

The Maddock Dune paleosol affects permeability and downward movement of water. Clay plus silt percentages are three- to four-fold higher in the 2Ab horizons than in overlying sands (Table 2). Hydraulic conductivity is about 35 times lower in the paleosol than in the overlying sands (Hopkins, 1997). The presence of calcium carbonate in the paleosol also contributes to its water-restrictive nature. This suggests that in addition to providing paleoenvironment and landscape evolution data, buried soils also exert a strong control on the hydrology of the modern Sandhills landscape by inducing lateral water flow or contributing to "perched" groundwater.

Historic soil erosion and landscape instability are more evident at the Soo Dune location than at the Maddock site, as shown by soil morphology and ^{14}C dating. The surface soil at the Soo Dune has less silt plus clay than the Maddock site, and the A horizon is considerably lower in soil organic matter (Table 2). The Soo Dune site has suffered more episodes of erosion and deposition as shown by its complex stratigraphy, which conforms well to the basic morphology of parabolic dunes (Figure 6). The radiocarbon ages of the uppermost and lowermost buried soils observed in the Soo Dune sequence are 580 +/- 90 BP and 890 +/- 50 BP, respectively (Running, 1997). The high sand percentage for the Soo Dune A horizon also implies relatively recent stabilization (Table 2). When Wanek (1964) did his field-work in the Sandhills, he found very high sand percentages (94.6 percent) in 30

Table 1. Physical and chemical properties of the Maddock Dune paleosol*

Horizon	Depth	Soil color	Reaction to HCl	Organic matter
	(cm)			(%)
A	0-8	10YR 2/1	-	-
A2	8-23	10YR 2/1	-	-
Bw	23-38	10YR 3/2	-	-
C1	53-69	10YR 4/3	-	-
C2	84-99	10YR 4/3	-	0.25
C3	114-130	10YR 4/3	-	0.30
C3	145-160	10YR 4/3	-	0.22
C4	160-175	10YR 4/3	-	0.25
C4	175-191	10YR 4/3	-	0.40
C4	191-206	10YR 4/3	-	0.51
2Ab	206-216	10YR 2/1.5	weakly effervescent	1.62
2Bw1	216-234	10YR 3/2	weakly effervescent	1.26
2B/C	234-239	2.5Y 6/3	weakly effervescent	0.64
3C	239-257	2.5Y 6/3	weakly effervescent	0.43
3C2	257-282	2.5Y 6/3	weakly effervescent	-

* Sampled in November, 1991; dashes indicate that data are not available.

Table 2. Soil texture, organic matter, and ^{14}C dates from the Maddock* and Soo Dune soils

Soil	Horizon	Depth	Sand	Silt	Clay	Organic matter	Conventional
		(cm)	—— (%) ——			(%)	radiocarbon age
Maddock	A1	0-6	86.7	8.7	4.6	4.17	-
	A2	6-36	95.0	2.8	2.2	0.88	-
	A3	36-60	96.4	1.8	1.8	0.29	-
	C2	114-156	96.8	1.7	1.5	0.17	-
	C3	156-178	95.7	2.7	1.6	0.17	-
	2A1b	178-210	87.7	8.2	4.1	1.15	2370 +/- 60 BP
	2A2b	210-216	86.7	6.6	6.7	1.87	-
	2C1	216-272	92.1	4.7	3.2	0.05	-
Soo Dune	A1	0-38	96.4	0.9	2.7	0.74	-
	A2	38-48	97.0	0.2	2.8	0.95	-
	C1	48-80	96.3	0.9	2.8	0.84	-
	Ab1	80-95	90.1	5.1	4.8	1.39	580 +/- 90 BP
	Ab2	124-144	95.9	1.0	3.1	0.99	890 +/- 50 BP
	C2	156-196	96.7	1.8	1.5	0.46	-

*Sampled in August, 1992, 5 m north of initial sampling.

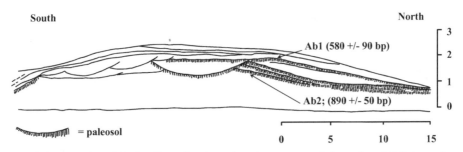

Figure 6. Cross section of the Soo Dune showing paleosol sequence and chronology (all distances in m, taken from Running, 1997).

upland sites. These high sand contents may reflect deposition of windblown sands during the "Dirty Thirties" that were stabilized by vegetation after better range management practice was initiated in 1941. The presence of fragile sedimentary structures indicative of eolian transport in the Soo Dune C1 horizon support this interpretation.

The physical and chemical properties of paleosols and their degree of development and age vary considerably between the Maddock and Soo Dune sites. They indicate a complex soil geomorphic record for this landscape. Despite the heterogeneity in characteristics of the paleosols, all are within the range of soil characteristics observed on the modern Sandhills landscape. There is no compelling reason to conclude that the paleosols formed in an environment unlike that of the present. The paleosol evidence suggests that the dunes have experienced numerous episodes of instability and soil development during Holocene time. Preliminary stable-carbon analysis of organic carbon in buried and modern soils at the Soo Dune also indicate that the paleosols formed under vegetation and climate conditions similar to the present (Running and Boutton 1996, Running 1997).

Methods and Materials

The landscape was mapped using a series of instrument points (IP) that were established on upland positions between the Soo Dunes and the Maddock paleosols and logged with GPS. Distances between these points were obtained with laser range finders and were generally about 145 m apart (Figure 4). Elevation was obtained with a laser level on 10 m intervals along the transect. Perpendicular transects were obtained 50 m north and south of alternate IPs to provide a more accurate description of relief in the study area. Vegetation data were obtained along the transect by noting key species identified by Nelson (1986) to map vegetation Habitat Type for the Sheyenne National Grasslands. The location of pond surfaces and their depth were also noted. A series of soil cores were dug with a barrel auger at IPs 3 and 4 and depth to paleosols were noted and elevations taken with a surveyor's level. Two paleosol samples were obtained for radiometric ^{14}C dating near IP 4. Standard radiometric dating was performed by Beta Analytic Inc.[1] Soil organic carbon was run by weight loss on ignition (NCR-13, 1998).

Results

Topography

The transect is oriented generally southeast from the origin point in Section 7 of Sandoun Township (T. 134 N., R. 53 W.) and extends well into Section 8 of the

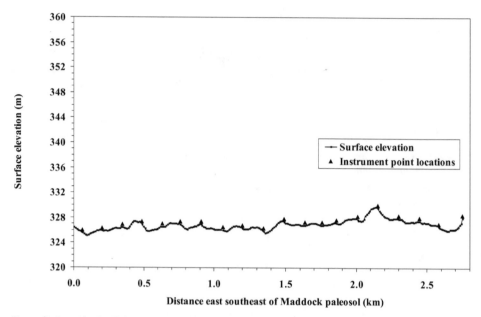

Figure 7. Smoothed relief on study transect from Maddock Dune paleosol to Soo Dune paleosol.

same Township (Figure 4). Range in elevation was just over 7 m, which is greater than normal for the area because the transect intersected a relatively high dune. Much of the landscape consists of lower undulating surfaces as shown by Figures 5 and 7. The entire transect is within the Hummocky Sandhills Association as mapped by Manske (1980), who stated that relief generally ranges between 1.5 and 3 m. Extremely choppy stabilized dunes were encountered between IPs 5 and 7, where drier conditions reduced density of grass and forb species. A larger dune was found between IPs 15 and 17 and bare ground was conspicuous in this vicinity (Figure 6). The variety of surface relief was clearly seen in the north-south transects (data not shown). Some show gentle slopes such as those at IPs 1, 12, 14, 18, and 20, while others were very choppy or irregular. Three of the 11 north-south transects had water ponded on one or both sides of the IP.

Paleosols at IP4

Two paleosols were observed below a relatively high, vegetated dune crest at IP 4 (Figure 8). The modern soil at this site is similar to that at the Maddock Dune (Table 1). It developed in clean, fine and medium sands extending to 2.56 m below the soil surface. From 2.56 to 3.18 m below the surface, multiple lenses of clean fine and medium sand including many thin (5-8 cm) brown (Munsell color 10YR 3.5/2) horizons were observed. These horizons were enriched in soil organic matter (i.e., 0.7 percent) relative to levels observed in eolian sands elsewhere on the study transect (see Table 1, C horizons). A rather thin black paleosol (Ab1) was observed 3.18-3.26 m below the soil surface that had 1.9 percent soil organic matter. This buried soil has a radiocarbon age of 1860 +/- 80 BP (Figure 8). Below the Ab1 horizon, the color remained black, but sand content increased. An older paleosol (Ab2) was observed from 3.45-3.58 m below the soil surface, which was very dark gray and had 0.9 percent soil organic matter. The Ab2 horizon has a

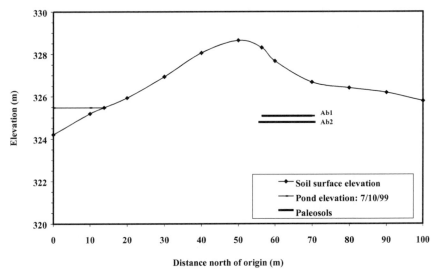

Figure 8. North-south cross section at IP4 showing pond elevation and position of paleosols. The ^{14}C radiometric date for the Ab1 is 1860 +/- 80 BP, and the Ab2 is 3170 +/- 110 BP.

radiocarbon age of 3170 +/- 110 BP. This ^{14}C date is the oldest recovered from the Sandhills and extends the existing Holocene pedologic record by 750 years. Above these two paleosols, clear contacts with clean medium sand were observed. These unconformities mark the shift from periods of dune stability, i.e. soil formation, to dune instability and erosion.

Vegetation

About half of the transect was classified as mixed grass prairie (Table 3). The low amount of water recorded reflects the fact that upland positions were specifically chosen for IPs to facilitate the elevation survey. Boundaries between Habitat Types were generally rather abrupt and conformed closely to elevation. For example, the sedge meadow Habitat Type was exclusively associated with shallow groundwater or ponded conditions as shown by Plate 6. Only 2 percent of the transect surface covered by water was in the tallgrass prairie Habitat Type, the remainder of the ponded water (98 percent) occupied the sedge meadow Habitat Type (Table 3). Manske (1980) confirmed the tight linkage between vegetation and topographic position observed by Shunk (1917), and stated that "the native habitat types are expressed in distinct belts on each hummock." In very choppy areas (IPs 5-8) in the current study, however, the boundary between tallgrass prairie and sedge meadow was more obscure; high willow (*Salix spp.*) density was observed in sedge meadows in this area due to excessive ponding.

Groundwater levels have changed markedly in the western Sandhills because of the wet climatic cycle the region is experiencing. Long-term monitoring of soil hydrology at the Maddock Dune wetland illustrates these changes. From 1990 to 1992, a truck was driven through the sedge meadow to service the groundwater observation well shown in Figure 5. In late July, 1990, the water table was 0.97 m below the soil surface at an elevation of 323.49 m. The wet cycle began in June, 1993, and the sedge meadow was transformed into a pond, which has persisted through each successive summer. The pond level was higher in July 1999, when we

Table 3. Distribution of vegetation Habitat Types on transect, representative plant species, and mean height above water table for each Habitat Type (open water found in sedge meadows)

Habitat Type	Percent	Common name	Genera and species	Mean height* above water table (m)
Dune complex	6.8	Sand bluestem	*Andropogon halli*	not determined
		Hairy grama	*Bouteloua hirsuta*	
		Cudweed sagewort	*Artemisia ludoviciana*	
		Bracted spiderwort	*Tradescantia bracteata*	
		Wild rose	*Rosa arkansana*	
Mixed-grass prairie	51.2	Blue grama	*Bouteloua gracilis*	1.66 (n=28)
		Prairie sandreed	*Calamovilfa longifolia*	
		Sun sedge	*Carex heliophila*	
		Needle-and-thread	*Stipa comata*	
		Green sagewort	*Artemisia dracunculus*	
Tallgrass prairie	22.2	Big bluestem	*Andropogon gerardi*	0.72 (n=45)
		Little bluestem	*Andropogon scoparius*	
		Indiangrass	*Sorghastrum nutans*	
		Switchgrass	*Panicum virgatum*	
		Kentucky bluegrass	*Poa pratensis*	
Sedge meadow	19.8	Wooly sedge	*Carex lanuginosa*	0.15 (n=29)
		Northern reedgrass	*Calamagrostis inexpansa*	
		Meadow anemone	*Anemone canadensis*	
		Panicled aster	*Aster simplex*	
		Willow	*Salix interior*	
Water	8.5			

*Data for IPs 1-5 only.

performed this fieldwork (325.15 m), than it has been since the wet cycle began in 1993. This represents an increase in water table elevation of 1.66 m during this 9-year interval, which is illustrated by Plates 7 and 8.

Discussion

Vegetation in the Sandhills can respond rather quickly to changes in depth to groundwater as shown from observations made during the last decade. Dense willow thickets several meters high have emerged from slight depressions on dune slopes replacing mixed grass prairie, which has apparently migrated upslope to drier conditions. Similarly, tallgrass prairie meadows have been transformed into sedge meadows replete with hydrophytic vegetation such as bulrush (*Scirpus spp.*) and willow (*Salix interior*). Some sedge meadows have evolved into semi-permanent open water ponds with bulrush, cattails (*Typha spp.*), and aquatic weeds, e.g. bladderwort (*Utricularia vulgaris*).

These striking changes in the proportion of vegetation Habitat Types were initiated by stepwise increases in growing season precipitation (May-October) from 1991–93 (Hopkins 1997). The increased precipitation allowed groundwater recovery after the severe 1980s drought. Summer and fall precipitation has remained higher than average through 1999 (Figure 9). In this case, the relationship between climatic variability and water table fluctuations is clear, direct, and rapid.

If the vegetation survey of the present study were performed in July 1990, when

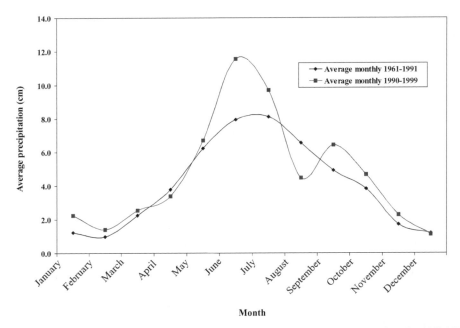

Figure 9. Average monthly precipitation from 1990-1999 at the study site compared to the 1961-1991 long-term monthly average at McLeod, ND.

the groundwater in the Maddock wetland was at an elevation of 323.49 m, the proportion of each vegetation Habitat Type would be considerably different. For example, the two large sedge meadows shown in Plate 6 are both contiguous with the extensive Maddock Dune wetland complex shown in Figure 5 and these zones would have been dominated by tallgrass prairie in 1990. The observations described above allow us to conclude that prairie vegetation in the Sandhills ecosystem exhibits a high degree of "plasticity" in order to adapt to changing groundwater elevations. We believe this is one aspect of biodiversity that improves our ability to evaluate the stratigraphic evidence for paleosol presence and preservation in the Sandhills with respect to emerging concepts of climatic change on the NGP during the Holocene Epoch.

Implications for Paleoenvironmental Interpretation

Evidence for post-glacial climatic variability across the NGP is myriad. A regional-scale model of Holocene climatic and environmental change is widely accepted (Vance and Last 1994; Last and Vance 1998); however, the magnitude of climate change and the response of specific environments to such changes remains equivocal. For example, a climatic reconstruction based on bio-salinity indicators at Moon Lake, 82 km northwest of the study transect, suggests that extended periods of severe drought (beyond the magnitude observed in the instrumental record) were more frequent before AD 1200 (Laird et al. 1996). These droughts are believed to have lasted on the temporal scale of centuries. If prolonged droughts were more frequent and severe before 1200 AD, preservation of buried soils in the fine sands at the study site would be limited at best. Erosion associated with such extensive episodes of landscape instability would presumably obliterate older soils and landforms. Instead, buried soils representing one third of Holocene time (3170 +/- 110 BP to the present) have been documented in this article. These dated paleosols could logically encompass dates of other paleosols that have been widely observed in the Sandhills (Wanek 1964; Burgess 1965; Bluemle 1979). The paleoenvironmental record preserved in the Sandhills does not appear to verify the interpretation presented by Laird et al. (1996) or, indeed, other researchers who identify the presence of buried soils in parabolic dunes as explicit evidence of past severe drought. What processes operative in the Sandhills landscape enabled soil formation to proceed through periods of climatic stress during the Holocene? Are these processes widespread and ongoing in other dune-dominated landscapes of the Great Plains? We believe a better understanding of the role of vegetation response to groundwater fluctuations, and therefore climatic variability, will help provide the answers.

Many studies of the paleoenvironmental record preserved in parabolic dunes on the NGP confirm that dune instability and erosion occur during periods characterized by relatively warm, dry climatic conditions; and dune stability and soil formation occur during cool, moist climatic periods. However, quantification of the climatic thresholds that trigger dune instability or stability remains unresolved. Many authors (Clayton et al. 1976; David 1971; David 1977) suggest dune instability is indicative of extreme drought in the past. A growing number of authors (Wolfe et al. 1994; Running 1996; Muhs et al. 1997; Running 1997), however, suggest that minor climatic fluctuations within the range of modern climatic variability explain episodes of stability/instability recorded in the region's parabolic dunes. We believe that groundwater fluctuations and vegetation response are a critical link in the indirect relationship between parabolic dune stability/instability and climate variability, and that this warrants additional research.

Since fine sands are easily entrained by winds, there are essentially only two factors that retard landscape erosion in such sediments. One of these critical factors is vegetation; the other is soil moisture content, which in the Sandhills translates to distance above the water table. Water under tension in soil pores resulting from either rainfall or capillary rise generates "apparent cohesion" in sands (Terzhagi and Peck 1948). The fact that wet sands do not erode led Baker (1967) to suggest the vast area of low relief northeast of McLeod resulted from complete deflation of deltaic sediments to the water table in the early Holocene. Similarly, paleosols can be preserved at lower landscape positions during extensive droughts because of capillary rise from the water table. Given a scenario of severe, recurrent Holocene droughts, one would expect that the majority of preserved paleosols were once within the range of the capillary fringe for fine and medium sands, i.e., they would have been wet soils; however, neither the Maddock nor Soo Dune paleosols exhibit morphologic indicators of wetness. It is also interesting to note that the paleosols observed at IP4 have essentially the same elevation as the water table in 1999 (Figure 8). If one assumes that the preservation of surface paleosols in sands requires a high water table, then our results can provide an approximate groundwater elevation for the period (3170 +/- 110 BP) when the paleosol at IP4 was protected from deflation. An alternative explanation might examine properties of the native Sandhills ecosystem that allow it to sustain drought with minimal damage.

Stabilizing Sands: The Role of Biodiversity in Landscape Evolution

Significant improvements in rangeland condition were accomplished after the Sheyenne Valley Grazing Association and the USFS initiated range improvement practices in the late 1930s. The number of blowouts on the Sheyenne National Grasslands was reduced from 270 to 50 during the 1940-80 period (USFS 1980). In only 40 years, soil formation has proceeded in those blowouts, slowly increasing organic matter content and creating better structured, thicker A horizons. The most critical factor in this process is vegetation colonization and succession on the bare sands, as suggested by Wanek (1964). The Sandhills ecosystem is adapted to utilize these barren sands; perhaps this attribute of biodiversity is responsible for rapid recovery during and after drought.

The "richest and most diverse flora" observed in the state of North Dakota was identified in Ransom, Richland, and Sargent Counties by the North Dakota State University Botany Department (Seiler and Barker 1985). Over 90 percent of the 871 taxa identified in the study can be found in the Sandhills proper, creating a remarkably rich and resilient ecosystem (Dr. William Barker, pers. comm.).

A distinct array of perennial vegetation, both grasses and forbs, colonize the blowouts in succession. Seeds of annual grasses and forbs typically exhibit viability for "extremely long periods" in harsh environments (Barbour et al. 1987) like the Sandhills blowouts. Burgess (1965) felt certain that "seeds and rhizomes of many species were present" in the bare sand. Some species such as Canada wildrye (*Elymus canadensis*) are found on the periphery, while other species, e.g., blowout grass (*Redfieldia flexuosa*) and sand sedge (*Cyperus schweinitzii*) are aggressive colonizers of the shifting sand. The seeds of blowout grass and sand bluestem can "germinate readily" when released, insuring higher cover for the revegetated blowout if conditions allow (Burgess 1965). Sieler and Barker (1985) identified 13 perennial forb species that colonize blowouts, and they indicate that two annual forbs, plains sunflower (*Helianthus petiolaris*) and winged pigweed (*Cycloloma atriplicifolium*) can "quickly cover the sand" during the growing season. Winged pigweed has a tumbleweed growth form and its stems can be "either erect or spreading," which would mitigate deflation losses in the healing blowouts (Great Plains Flora Association 1986). With time, sand bluestem (*Andropogon hallii*) spreads through the blowout (Barker and Sieler 1985), and its strongly rhizomatous nature fosters a tighter root mat (Great Plains Flora Association 1986). Wanek (1964) also found that mosses were common colonizers of bare sands in blowouts that were ungrazed.

The intrinsic diversity and adaptedness of the prairie vegetation, coupled with a complex seed bank, can enable both resistance and recovery to environmental stress. Dormancy is induced in numerous prairie species due to water deficits (Burgess 1965), enabling some species to be resistant to drought. The rush family (*Juncaceae*) is represented by 5 species in the Sandhills ranging from moist to dry soil water regimes (Sieler and Barker 1985). Big bluestem, a dominant component of the tallgrass prairie, exhibited more vigorous growth, as measured by culm height and leaf length, on sandy soils compared to finer-textures soils (Wali et al. 1973): this would favor resistance of this species in adverse conditions.

Conclusions

Pedologic evidence demonstrates that landscape instability is part and parcel of the Sandhills ecosystem. Early settlers of the Sandhills were not aware of the paleoenvironmental record that lay beneath their feet; nor were they aware of the devastating effects that prolonged overgrazing would have on the inherently resilient prairie. Clearly, the dune-dominated Sandhills are characterized by considerable ecological and stratigraphic complexity even across short distances. Many intervals of soil erosion and dune instability, and intervening intervals of dune stability and soil formation have occurred since 3170 +/- 110 BP. We propose that the oldest paleosol in this article, that found at IP4, be named the McLeod Surface paleosol. There are likely other paleosols in the Sandhills, from early- and middle-Holocene time that have not yet been discovered or dated. Future research will be done on an area basis to evaluate the three-dimensional extent of the McLeod Surface and other Holocene paleosols near the linear transect established in this article.

This article demonstrates that significant changes in vegetation Habitat Type distribution have occurred since the drought of the late 1980s. Two important observations can be made. First, vegetation adapted to the Sandhills environment is resilient and capable of adjusting to rapid, annual and decadal, fluctuations in the environment. The native prairie vegetation appears to be well poised for environmental stress, and it can easily adapt to either drier or wetter conditions. Perhaps the resilience of vegetation in the Sandhills make this landscape less fragile, under natural conditions, than has been previously recognized. The Sandhills landscape may act as a refugia and seed bank during periods of regional-scale severe drought because of its tremendous species richness (Seiler and Barker, 1985). Secondly, it is clear that changing groundwater elevations have a profound impact on the distribution of vegetation Habitat Types and, presumably, on overall measures of species composition and vigor. Decisions concerning future land use in the Sandhills must acknowledge these principles as human activities that lower water table elevations will affect distribution of native vegetation. If groundwater drawdown is sufficient and vegetation thresholds are not known, some loss of vegetation cover is to be expected. If vegetation cover is lost, commensurate increases in soil erosion and dune instability would occur. Therefore, while the Sheyenne Delta aquifer can supply a vast amount of water (Armstrong, 1982), extraction of groundwater for irrigation and municipal purposes must be carefully monitored. Native prairie should be preserved, where possible, and sustainably grazed. Producers developing irrigation capabilities need to select and utilize rapidly growing cover crops to prevent erosion and maintain their soil productivity. Sustainable grazing by livestock is the single land use practice that most closely approaches the pre-settlement conditions that fostered the plant biodiversity observed today.

Acknowledgements

We would like to thank Mark Aurit, Amy Landis and Tim Morrell, students from the Department of Geography at the University of Wisconsin-Eau Claire (UWEC), and Andrea Travnicek, Natural Resources Management student from North Dakota State University, for their assistance in fieldwork. The Office of University Research and the Department of Geography at UWEC funded the UWEC contingent in the field. The Natural Resources Conservation Service Wet Soils Monitoring Program funded hydrological and soil characterization at the Maddock Dune site; and the Gunlogson Fund of the North Dakota Institute for Regional Studies supported ^{14}C dating for this article. We appreciate the support of Mr. Bryan Stotts, District Ranger of the Sheyenne National Grasslands, for permission to conduct the survey. We would like to express our sincere thanks to Mr. Bob Berg, who graciously loaned us his ATV when ours was "kaput," and supplied gas and mobile tire repair service. The continued and competent assistance of Mr. Lynn Foss in hydrological monitoring is much appreciated.

Notes

1. Beta Analytic Inc. University Branch, 4985 S.W. 74 Court, Miami FL 33155. Use of vendor name does not imply endorsement by the North Dakota Agricultural Experiment Station or the University of Wisconsin-Eau Claire.

Plate 9: See Smith and Radenbaugh

Plate 10: See Paul

Plate 11: see Paul

Plate 12: see Paul

Plate 13: see Paul

Plate 14: see Paul

Plate 15: see Paul

CHAPTER 4

ECOLOGY AND ECOSYSTEM FUNCTIONS OF NATIVE PRAIRIE AND TAME GRASSLANDS IN THE NORTHERN GREAT PLAINS

Duane A. Peltzer

ABSTRACT. Grasslands occupy more than a quarter of the world's land surface area and are valuable sources of food, forage, fibre and medicine. Grasslands also provide a wide range of critical ecosystem services (i.e. functions important to people) such as air and water purification, nutrient cycling, regulation of atmospheric gases, and are an important source of biodiversity. Despite the extent and widespread use of grasslands worldwide, the effects of human activities on grassland ecosystem functions by the removal of native herbivores, agriculture, habitat fragmentation and the introduction of exotic plants are poorly understood. Here, I compare the ecology and ecosystem functions provided by native prairie and tame grasslands dominated by exotic species using examples from the northern Great Plains. I suggest that one powerful approach to prairie conservation is to recognize and value the important ecosystem services that these grasslands provide.

Introduction

Grasslands and pastures cover 68 million km², or about one quarter of the Earth's land surface (Shantz 1954, Meyer and Turner 1992). In North America, native grasslands are the largest vegetation biome on the continent, and before European settlement covered some 162 million hectares (ha) (Sampson and Knopf 1994). Temperate grasslands such as those found at mid-latitudes in western Canada and the United States are used extensively for forage or, where precipitation and soils permit, the cultivation of crops (Coupland 1979). Many native grasslands have been converted to agricultural lands during the past century. For example, in western Canada only 1-20 percent of native prairie remains including: <20 percent of mixedgrass prairie, <5 percent of fescue prairie, and <1 percent of tallgrass prairie (Samson and Knopf 1994, Morgan et al. 1995).

Marginal agricultural lands have been planted with exotic perennial plant species during droughts, or when government policy and economic conditions are favourable. In Canada, millions of hectares of marginal cropland were planted to tame perennial grasses during and after the drought of the 1930s (Gray 1996), more recently as forage, and in the United States as part of the Conservation Reserve Program (CRP) (Skold 1989).

In this article I review some of the causes and consequences of the extensive sodbusting of northern native prairie. I also suggest that native prairie provides many important ecosystem services, such as nutrient cycling and regulation of atmospheric gases. Furthermore, the services provided by native grasslands may be superior to those provided by tame grasslands, that is, grasslands dominated by exotic or improved varieties of grass species. By valuing and conserving ecosystem services in native grasslands, we also explicitly recognize that local and regional land management can affect global health (Rapport et al. 1998).

I first outline the perceived value of prairie by European settlers in order to better understand the underlying reasons for the rapid and extensive conversion of native grasslands to agricultural lands, as well as the expansion of tame grasslands in Canada. Next, the ecology of native prairie and tame grasslands, and the ecological consequences of converting native prairie to other land uses are discussed. The concept of ecosystem services (*sensu* Daily 1997, Chapin et al. 1997) is then reviewed and contrasted between native prairie and tame grasslands. Lastly, I suggest what future research is needed to address the problem of conserving ecosystem services in grasslands.

Human Values and Perceptions of Grasslands in the Northern Great Plains

Native prairie in Canada was viewed by European settlers as a vast, lonely, and empty land. One early written account of this region is given by Captain John Palliser, an early surveyor in western Canada. He described the region around southeastern Alberta and southwestern Saskatchewan — which is now called Palliser's triangle — as "unfit for settlement" (Potyondi 1995). This is still a contemporary view. For example, in Wallace Stegner's classic book *Wolf Willow* (1962), prairie is variously described as vast, lonely, geometrical, flat, stable and empty. Similarly, prairie is usually perceived as lacking in resources, as having too few trees, too little water, and as being either too cold or too hot for many crops (Manning 1995, Olson 1995).

Prior to the introduction of farming in the early 1900s, the only major source of economic activity in western Canada was through cattle ranching (Potyondi 1995). Today, the main economic value of grasslands is agriculture, followed by ranching (SERM 1997). The success of these two dominant uses of the northern prairie region — agriculture and forage — depends on the natural environment (such as precipitation, soils) and on economic conditions (such as market prices, crop insurance subsidies) (Coupland 1979). For example, satellite images of the Alberta-Montana border show extensive croplands on the American side, but native rangelands in Canada in several areas. Different land uses in this region are not due to environmental constraints *per se*, but to differences in crop insurance provided by the United States and Canadian governments (see Figure 1 in Dormaar and Smoliak 1985).

Wresting an economic livelihood from prairie land was a perspective unique to European settlers. Aldo Leopold (1949: 188) recognized this view more than fifty years ago in the American Midwest: "To the labourer in the sweat of his labour, the raw stuff on his anvil is an adversary to be conquered. So was wilderness an adversary to the pioneer." Prairie was viewed by European settlers and their contemporaries as a vast, empty, stable place where a livelihood had to be made from an environment with many deficiencies. Thus, the northern prairies have not been highly valued as agricultural lands or tame pastures, partly because of these views (Joern and Keeler 1995, Potayandi 1995).

Ecology of Native Prairie and Tame Grasslands

The distribution and species composition of native prairie is shaped strongly by climate and soils. Temperate grasslands are found at mid-latitudes in continental climates (Archibold 1995). For example, the northern Great Plains have a large annual range in temperature (mean annual temperatures of 1–7°C with extremes from -43°C to +38°C), 150–500 mm of annual precipitation, and frequent droughts (Risser 1988, Environment Canada 1993). Soils in the northern Great Plains are formed by ancient sediments, glacial deposits and the decomposition of prairie plants over millennia (Jenny 1980, Pielou 1991, Agriculture Canada 1992).

The northern Great Plains contain prairie, a relatively new flora, in geological terms, which has attained its present distribution only during the past 10,000 years, i.e. since the last glaciation (Axelrod 1985, Pielou 1991). Prairie originally covered about 350 million ha in North America (Archibold 1995); but since the mid-1800s much of this region has been converted to agriculture or planted to tame pastures (Sampson and Knopf 1994). Prairie vegetation is usually dominated by long-lived perennial grasses and forbs, and contains fewer woody and annual plants than other ecosystems. Grasses typically account for only about 20 percent of the total plant species, but for about 90 percent of the biomass at any site (Coupland 1979, French 1979). The growth habits (i.e. height) and species composition (i.e. photosynthetic types, C_3 vs. C_4) of dominant grasses are used to classify prairie as shortgrass (<0.3 m tall), mixedgrass (0.3–1.2 m) or tallgrass (ca. 2 m). Most northern prairie is mixedgrass; but there are small areas of tallgrass prairie in southwestern Manitoba and North Dakota (Scott 1995). There are several excellent reviews of the vegetation, history, distribution and ecology of prairie in North America (Coupland 1950, Weaver 1954, Weaver and Albertson 1956, Coupland 1979, French 1979, Archibold and Wilson 1981, Axelrod 1985, Archibold 1995, Joern and Keeler 1995, Samson and Knopf 1996, Knapp et al. 1997).

Most life on the prairie is belowground; up to 80 percent of plant biomass is roots and rhizomes. In addition, nearly all of the decomposing shoots, roots and animals eventually make their way to the soil community (Stanton 1988). For example, the top 30 cm (12") of soil at Matador, Saskatchewan contains 100 percent of the plant rhizomes, 98 percent of the litter, and 55 percent of the roots (Coupland 1979). The aboveground plant mass added each year in mixedgrass prairie is 90–200 g/m^2 whereas about 1,400 g/m^2 community biomass is added belowground in only the top 30 cm of soil (Coupland 1979). Hence, the dominant energy flows are found within the plant-soil subsystem and not in the canopy (as summarized in Figure 1).

Allocation to belowground production is much lower in tame grasslands dominated by crested wheatgrass (*Agropyron cristatum*) than in native prairie (Christian and Wilson 1999). Pastures planted to either crested wheatgrass or smooth brome (*Bromus inermis*) allocate less biomass belowground than native grasses, resulting in lower root mass and reduced quality of organic materials to the soil (Redente et al. 1989, Lesica and DeLuca 1996). Although the relatively high shoot mass of tame grasses is favoured for forage, these grasses may not necessarily outperform native species over the long term. Tame forage species are usually planted in previously cultivated fields, and then their aboveground production is compared to uncultivated native prairie: this means that the effects of cultivation are often confounded with increased plant production of tame species (Coupland 1979).

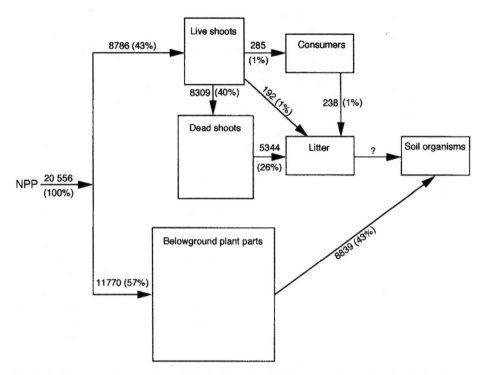

Figure 1. Mean anual energy flow (kJ/m²) through mixed-grass prairie at Matador Saskatchewan, Canada. Box size represents the energy content of that compartment in kJ/m². NPP = Net Primary Production i.e. net energy capture by prairie plants over a year. (redrawn and adapted from Coupland 1979).

Several ecological processes including fire, grazing, and drought are intimately associated with grasslands. Fire has long been considered a major force which shaped and sustained grasslands (Stewart 1955, Rowe 1969, Wright and Bailey 1982, Collins et al. 1998). Fire removes accumulated litter, speeds nutrient cycling, and can kill woody plant species, preventing or slowing their growth or invasion into grasslands (Wright and Bailey 1982, Knapp and Seastedt 1988). Large grazers such as bison (*Bison bison*) are attracted to recent burns, and repeatedly graze these patches of grasslands (Knapp et al. 1999). Grazers also speed the recycling of nutrients, create small-scale disturbances by grazing, trampling and wallowing, and may preferentially eat the dominant grasses and release subdominant species such as forbs. All of these factors increase plant diversity (Huntley 1991, Gibson et al. 1993, Collins et al. 1998). Many of the plant traits conferring resistance to grazing are also adaptations to fire, drought or seasonal water shortages that characterize grassland regions (Cougenhour 1985). For example, prairie plants have relatively dense tissues, growing meristems located close to the ground, and deep, extensive root systems which allow these plants to thrive under climatic stress, frequent fires and repeated grazing, making prairie resilient to these kinds of disturbances. Nevertheless, with agricultural disturbances and the introduction of exotic grassland plants and weeds, prairie can be converted to relatively stable, tame grasslands (Dormaar et al. 1995, Peltzer et al. unpublished data).

Management of Grasslands

Grasslands in the northern Great Plains may have been managed for millennia. For example, fire was used extensively by humans for hunting, creating pasture for game species, decreasing the abundance of woody plants, increasing the abundance of food plants (e.g. seeds, culativated maize), and possibly as a weapon to rout enemies (Stewart 1955, Gilmore 1977, Doebley 1984, Kindscher 1992). Although the extent to which prairie is the result of anthropogenic or natural fires is uncertain, there is no doubt that prairie and fire are intimately related (Rowe 1969, Wright and Bailey 1982).

With European settlement, a shift occurred on the northern Great Plains away from an economy based on bison (ca. 1850–1879) to one based on ranching (1880–1920). The current system dominated by agriculture began with the land rush of 1908 (Johnston 1970): settlement resulted in the concept of land ownership and fencing of property. This indirectly resulted in the deterioration of range conditions caused by the overgrazing of fenced pastures. The entire process of homesteading, sodbusting, cultivating and land abandonment occurred from 1908 to the 1930s (Spector 1983, Potyondi 1995).

Early in the 1900s, the provincial governments of the Canadian Prairies had a policy of sodbreaking in place. For example, the Saskatchewan Department of Agriculture suggested as early as 1909, and continuing into the 1920s, that "the first task of the farmer is to destroy native plants in order to prepare a place for cultivated plants to grow" (Potyondi 1995: 112). Thus, policy both established and reinforced the view that land should be valued in terms of its economic contribution. In Saskatchewan, agricultural development flourished under these policies despite a series of droughts in 1918 and the 1920s, so that by 1931 about 60 percent of southern Saskatchewan was cultivated. Today about 70 percent is cultivated while only 18 percent is in native prairie, and 12 percent is in "improved" grasslands (Statistics Canada 2000). Until recently, bringing prairie into cultivation was considered "improving the land" (Potyondi 1995, Lesica and DeLuca 1996); unfortunately, disturbed prairie may never be regained (Dormaar et al. 1995, Kindscher and Tieszen 1998, Bakker and Berendse 1999; discussed below). There is no doubt that the short-term economic output of many regions was enhanced by switching from ranching to agriculture; but the land itself was not "improved" in this process.

Cultivating soils results in the net loss of major elements such as carbon and nitrogen through increased soil exposure to wind and soil erosion, increased mineralization, mixing surface soil with deeper strata, and enhancing decomposition rates (Jenny 1980, Burke et al. 1989, 1995). For example, about 18–30 percent of soil organic matter (SOM), the organic carbon that maintains soil structure and fertility, is lost during the first five years of cultivation (Burke et al. 1995); and up to 75 percent of SOM can be lost with long-term cultivation (Buyanovsky and Wagner 1998). In some areas, SOM may continue to decline beyond 90 years of cultivation because of continued erosional losses (Tiessen et al. 1982, 1994). Modern agricultural techniques (e.g. no till) can reduce or reverse losses of soil carbon and nitrogen due to cultivation (Buyanovsky and Wagner 1998); but where alternative management does not improve the condition of soil on marginal agricultural lands, other approaches are needed. For example, the United States Conservation Reserve Program (CRP), initiated in the 1980s, promoted soil conservation by paying farmers to plant marginal agricultural lands with tame perennial grasses (Skold 1989). Unfortunately, some landowners have cultivated native prairie at the same time they planted marginal cropland to grasses as part of the CRP: thus, the CRP indirectly subsidized the breaking of prairie (Lesica 1995). The impact of changing land use, however, goes far beyond altered soil fertility and nutrient conservation.

Today, little of the original prairie remains across western Canada and the United States (see Table 1 in Samson and Knopf 1994). Agriculture has destroyed and fragmented native prairie, leaving most remaining prairie surrounded by agricultural fields, marginal lands or roads; and it has indirectly resulted in the extirpation of many species populations (see general discussions in Joern and Keeler 1995, Samson and Knopf 1996).

Small fragments of remnant native prairie are susceptible to rapid losses of plant species. For example, Leach and Givnish (1996) found that prairie fragments lost 8–60 percent of their original plant species over a 32–52 year period in Wisconsin. Short (<25 cm tall), small-seeded and nitrogen-fixing species had the highest rates of extirpation. These losses are presumably caused by the disruption of large-scale processes such as fire or grazing. Similarly, the diversity of invertebrates (e.g. beetles and spiders) declines significantly in smaller patches of remnant mixedgrass prairie in Saskatchewan (Pepper 1999). In contrast, archipelagos of prairie patches retain more plant species than single, large isolated patches of a similar size, suggesting that patch size *per se* does not constrain total species richness for plants (Simberloff and Gotelli 1984). Clearly, more research is needed to understand the long-term consequences of habitat loss and fragmentation in prairies.

Why should the loss of native prairie species be of concern? The most pressing concerns, summarized in recent reviews by Samson and Knopf (1994) and Joern and Keeler (1995), are: 1) losses of native prairie are typically 82–99 percent, much greater than losses in other major North American ecosystems including remnant old growth forests; and 2) more than one-third of species considered endangered by the Committee on the Status of Endangered Wildlife in Canada are found in native prairie (PCAP 1998). Thus, although prairie is valuable in terms of supporting economic activities, it is also a rich source of biodiversity that has been highly impacted by changing land use. In addition, prairie is also threatened by exotic plant species introduced by humans either intentionally for forage or soil conservation, or unintentionally as invasive weeds.

The Introduction of Exotic Species

Tame (exotic) plant species cover millions of hectares across the Great Plains. Some of these introductions are intentional: for example, forage grasses (*Agropyron cristatum, Bromus inermis*) and forbs (*Medicago sativa, Melilotus* spp). Crested wheatgrass (*A. cristatum*) was planted extensively during the 1930s on the Canadian Prairies to control soil erosion (Grey 1996), and more recently as forage for cattle because it is drought-resistant, cold-tolerant, easy to seed, and is a relatively productive forage species (Rogler and Lorenz 1983, Gray 1996). There can be economic gain in planting exotic species as tame pastures (Vallentine 1989), but species introductions can also have enormous environmental costs. For example, in the United States, the 79 most harmful alien species caused $97 billion (US) in economic losses from 1906 to 1991 (Stein and Flack 1996).

Unintentional introductions cause the degradation of millions of hectares of native prairie in Canada and the United States (Mack 1981, 1989, White et al. 1993). Some of the worst introduced weeds in western Canada include leafy spurge (*Euphorbia esula*) and spotted knapweed (*Centaurea maculosa*) (Best et al. 1980s). *Euphorbia esula* causes economic losses of 50–75 percent due to reduced productivity of grazing; it also lowers the diversity of native plant species and diminishes habitat quality for wildlife (Selleck et al. 1962, Belcher and Wilson 1989).

Control of exotic plants can be extremely difficult. For example, Lym and Messersmith (1985) examined several methods of controlling *Euphorbia* using different herbicides, and found that the most effective treatments had only about 89 percent control after two years of repeated applications. These authors suggest it may take 5–10 years of annual herbicide applications to eradicate a stand of *Euphorbia*, an incredibly intensive and costly effort. Some species originally introduced for forage have subsequently become invasive: for example, smooth brome grass (*Bromus inermis*) is widely planted as a tame forage species, but is able to invade along moist draws, roadside ditches and riparian zones. Control of invading *Bromus inermis* is difficult; only intensive, repeated herbicide applications coupled with spring burns can significantly reduce or eliminate established patches of brome (Grilz and Romo 1995). Exotic plant species can also have long-term impacts on grassland ecosystems.

Long-term Effects of Introduced Species

There are important long-term consequences of introducing exotic plants over large areas of the Great Plains (Dormaar and Smoliak 1985, Lesica and DeLuca 1996, Christian and Wilson 1999). Previously cultivated fields planted to either

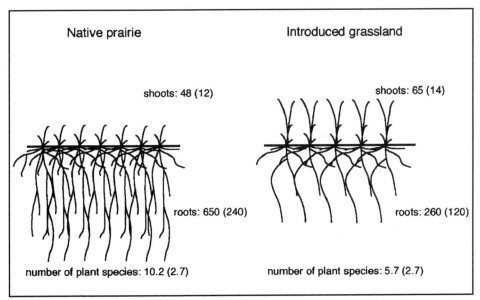

Figure 2. Comparison of biomass and plant species richness in native prairie and introduced grasslands in Saskatchewan, Canada. Numbers represent the mean peak dry weight of peak plant biomass in g/m² (100 g/m² = 1 t/ha); numbers in brackets are 1 SD of the mean. Root biomass is for the top 10 cm of soil only. Data are taken from Christian and Wilson's (1999) study in Grasslands National Park, Saskatchewan, Canada.

crested wheatgrass or Russian wild rye (*Elymus junceus*) have lower root mass than native plants, and cannot restore or maintain the chemical quality of soils relative to native prairie (Dormaar et al. 1995). Planted stands of tame forage species usually remain as monocultures for decades or longer after planting (Romo et al. 1994, Christian and Wilson 1999). In general, biodiversity of many kinds of organisms (e.g. songbirds, insects, small mammals) is much lower in tame grasslands than in native prairie (see Belcher and Wilson 1989, Sutter and Brigham 1998, Christian and Wilson 1999, Davis and Duncan 1999, Pepper 1999).

One of the best known and widely planted tame forage species in North American rangelands is crested wheatgrass (*Agropyron cristatum* (including *A. desertorum*). A recent study in Grasslands National Park (southwestern Saskatchewan) found that old-fields planted to *Agropyron* have about 25 percent lower soil carbon than old-fields allowed to undergo succession to native prairie species (Christian and Wilson 1999; see Figure 2). In addition *Agropyron* contributes less SOM of lower quality to soils, resulting in poorer soil structure and reduced biological activity belowground compared to native prairie (Dormaar et al. 1978). These studies suggest that planting tame pasture species moves carbon from the soil into the atmosphere, possibly increasing levels of greenhouse gases. For example, *Agropyron* pastures in southern Saskatchewan may have released $3.3–4.8 \times 10^{14}$ g of carbon to the atmosphere after about fifty years of succession that would otherwise be stored in the soil under native prairie (Christian and Wilson 1999).

Stability

Stability is a complex phenomenon consisting of resilience and resistance components. Resilience is the time it takes for a community to return to its previous state whereas resistance is the force or energy needed to initiate change in the community (Holling 1973, Peterson et al. 1998). One of the best documented examples of stability in grasslands is the perturbation and recovery of prairie during the extensive drought of the 1930s: across the western United States, range condition deteriorated and tallgrass prairie was replaced by mixed- or shortgrass prairie species. With the return of moister weather in the early 1940s, prairie vegetation recovered to its pre-drought composition (Weaver 1943, 1954).

Stability of native prairie and tame grasslands can be measured as the variability in aboveground production (e.g. using the coefficient of variation (CV) or elasticity of production) with changes in precipitation or climate (Noy-Meir and Walker 1986). Noy-Meir and Walker (1986, their Table 1) found that "improved [i.e. tame] pastures" in Israel's semiarid grasslands have higher CVs of annual aboveground production and are more sensitive to fluctuations in rainfall than native grasslands; similar tests of stability between native prairie and tame grasslands have not been done in North America. Nevertheless, stability is of concern for grassland management because degradation of range condition can result in sudden, unexpected and discontinuous changes in plant community composition (Friedel 1991, Laycock 1991). The current model of range management is based on the climax model of Dyksterhuis (1949, 1958). This model assumes that reducing grazing pressure will improve range condition, but this assumption does not always hold. For example, if intense grazing causes a shift in the plant community to a new stable state, subsequent reductions of grazing pressure may not return the community to its original state. Such shifts have occurred in the southern United States where overgrazing has caused irreversible shifts from formerly productive native grass-lands to arid shrublands (Schlesinger et al. 1990).

In summary, prairies are resilient because they can withstand drought, fire and grazing pressures; but after sufficient disturbance, prairie can be driven to a new stable state through intense grazing, soil disturbance or the introduction of exotic species. This idea was presaged by Leopold (1949: 218), who wrote that "man-made changes are of a different order than evolutionary changes, and have effects more comprehensive than is intended or foreseen." Clearly there are fundamental changes in ecological processes as a result of converting prairie to cultivated land and tame pasture species. In the next section, I discuss possible changes in eco-logical functions provided by native prairie and tame grassland ecosystems.

What are Ecosystem Functions and How are They Provided by Grasslands?

Many functions supplied to humans by grassland ecosystems are taken for granted: soil fertility, drought resistance, water purification, carbon storage, nutri-ent cycling, decomposition, aesthetics, biodiversity, and climate regulation to name but a few (Daily 1997). If we try to attach an economic value to these ecosystem functions for the entire earth, it totals at least twice the economic activity of the entire world each year (Costanza et al. 1997). By thinking of ecological functions as services to humans, we explicitly recognize the non-market values of natural sys-tems. In addition, these services often are the least well understood but possibly the most important value of biodiversity (Westman 1977, Walker 1992, 1995, West 1993). Many of the ecosystem services provided by grasslands are summarized in Table 1 (see also discussion by Sala and Paruelo 1997). In general, most ecosystem functions are likely higher in native prairie than in tame grasslands; but much more work is needed to compare the functions of native and tame grasslands.

Many ecosystem services provided by grasslands are belowground and thus inti-mately associated with soils (Stanton 1988, Sala and Paruelo 1997). Because soils are either lost through erosion or degraded by continued cultivation or the intro-duction of exotic plants, losses of ecosystem functions should also occur (Pimental et al. 1995, Daily et al. 1997). For example, SOM in soils under exotic grasses is lower than under native prairie (Dormaar et al. 1995, Christian and Wilson 1999). Soils in fields abandoned from cultivation for 50 years in northeastern Colorado were similar to native prairie in many respects except for persistent reductions in levels of SOM and silt content (Burke et al. 1995). This suggests that 50 years may be adequate for recovery of active SOM fractions and nutrient availability, but recovery of total SOM will take much longer. Similarly, the quantity, quality and sta-bility of SOM improved in fields abandoned from cultivation in 1925, 1927, or 1950 and allowed to undergo succession to native plants in southern Alberta; however, complete recovery of SOM may take 75–150 years (Dormaar et al. 1990, Kindscher and Tieszen 1998). Because SOM is critical in determining the rates of nutrient cycles, carbon storage capacity, water holding capacity (Witkamp 1971, Jenny 1980), losses in quantity and quality of SOM may also reduce soil function (Daily 1997).

Most prairie is belowground and thus unseen: hence long-term reductions in soil quality, productivity and ecosystem functions related to soils are really "hidden values" of native prairie. How can we quantify changes in ecosystem functions between different land-uses such as native prairie vs. tame grasslands?

Table 1: Comparison of Some Ecosystem Functions Provided by Native Prairieand Introduced Grasslands in the Northern Great Plains*

Ecosystem Service or Function	Native Prairie	Introduced Grasslands
Resistance to Disturbance		
Fire	High	Varies
Drought	High	High
Grazing	High	High
Soil		
Maintenance and fertility	High	Low
Formation	High	Low
Erosion	Low	Varies
Decomposition of organic materials Varies?		High
Cycling of nutrients	Varies	High?
Water		
Purification	High?	?
Storage and retention	High	Varies
Mitigation of drought or floods	High	Varies
Regulation of Atmospheric Gases	High?	Varies?
Moderation of Weather Extremes	High?	High
The Biota		
Aboveground productivity	Varies	High
Belowground productivity	High	Low
Biodiversity	High	Low
Habitat or refuge for species	High	Low
Maintenance of biodiversity	High	Low
Pollination	High	Low
Food Production	Low	Varies
Commercial/Economic Value	Varies/Low	Varies/High
Recreation	Varies/High	Low
Cultural Value	High	Low

* Data are summarized from Coupland 1979, Dormaar et al. 1995, Lesica and DeLuca 1996, Costanza et al. 1997, Daily 1997, Sala and Paruelo 1997, Christian and Wilson 1999."?" refers to functions which are not explicitly addressed in the literature, but the presumed value is given.

Quantifying Ecosystem-level Changes

Changes in ecosystem functions can be examined in several ways. For example, we can measure the symptoms associated with ecosystem stress, collectively termed an "ecosystem distress syndrome" (EDS), where distress is defined as impairment of, or reduction in, one or more ecosystem functions (Rapport et al. 1998). These symptoms (e.g. reduced productivity, increased respiration, presence of exotic species) can be used to indicate that declines in ecosystem health or functions. Here, I focus on differences in ecosystem functions between native prairie and tame grasslands; hence methods assessing changes in land-use or management are particularly useful.

Changes in land-use patterns (e.g. extent of cultivation) can be quantified using government records, aerial photography, or remote sensing of land cover. Large-scale changes in species composition and impacts on ecosystem function can be examined in several ways (reviewed by Vogt et al. 1997). Aerial photography can be used to distinguish weeds or invasive exotic plants in some native communities (Everitt et al. 1989). On regional scales, remote sensing coupled with geographic information systems (GIS) can be used to map patterns of species diversity in the landscape (e.g. Lauver 1997). Many complementary techniques can be used simultaneously to provide a richer understanding of changes in grassland functions. For example, Stohlgren et al. (1998) provided several lines of evidence including models, hydrological records, and vegetation data to show that the effects of land-use practices on regional climate may overshadow larger-scale temperature changes associated with increased greenhouse gases. Regional monitoring of land-use changes is appropriate for answering questions of habitat availability and fragmentation; however, changes in specific ecosystem functions should be monitored in more intensive follow-up surveys which are more appropriate for determining the effects of different land management regimes.

One powerful approach to understanding variation in ecosystem functions is to survey a variety of native and tame grasslands subjected to different kinds of management regimes. We can then compare ecosystem functions in native prairie vs. tame grasslands using the methods described above. Here, native prairie acts as a control, or baseline data, against which we can compare the effects of different management regimes. Leopold (1949: 196) stated that "a science of land health needs a base datum of normality, a picture of how healthy land maintains itself." For example, prairie grasslands can be superior carbon sinks relative to agricultural lands, forests or tame grasslands because they are better able to modulate greenhouse gases (e.g. CO_2) (Seastedt and Knapp 1993, Houghton et al. 1998). This kind of information is critical to understanding the long-term consequences of changing land use and land management.

In summary, there are several tools that can be used to quantify and monitor changes in land use and associated ecosystem functions; however, the use and function of grassland ecosystems are ultimately determined by socio-economic conditions (Holling 1986, Pearson and Ison 1997). In grasslands, priority should go to maintaining soils and their associated ecosystem functions (West 1993); but how can this be accomplished?

The Future of Grasslands

Conservation and Management of Native Prairie

We can take two approaches to conserving grasslands and their ecosystem functions: we can conserve remaining areas of native prairie; or we can try to restore functions to degraded lands by restoring prairie or planting tame grasslands. Conservation of native prairie can take many forms such as creating protected areas, tax easements on private land, moderate grazing of native rangelands or prescribed burning (PCAP 1998). Even within protected areas of native prairie, there are risks of losing species because of habitat fragmentation or altered grazing and fire regimes (discussed above). As a result, most conservation efforts will require active management of species composition in plant communities within remnant areas of prairie, in the surrounding buffer zones, and through prairie restorations (e.g. Hobbs and Huenneke 1992, Peltzer et al. 2000).

Active management for the conservation of remnant prairies may involve eliminating exotic species, enhancing the quality of buffer zones surrounding the area, and using fire and grazing to maintain an appropriate disturbance regime. Invasive exotic species are an immediate threat to the long-term species composition and productivity of native prairie, and their elimination should be the highest priority for conservation efforts. Similarly, preventing the invasion of remnant prairies by invasive species is necessary. One of the simplest ways to prevent invasions may be to improve the species composition of buffer zones surrounding remnant prairies, for example by eliminating weedy species in road ditches surrounding patches of native prairie.

Little information is available on the long-term effects of using fire and grazing together to maintain the species composition of prairie. Recent work in Kansas tallgrass prairie (Collins et al. 1998, Knapp et al. 1999) shows that repeated burning reduces plant diversity and alters community composition, but that grazing by bison counteracts the effects of fire by reducing the abundance of dominant grasses and maintaining a diverse plant community. More work is needed in the Canadian prairie region to determine what combinations of fire and grazing are needed to maintain the long-term species composition and productivity of prairie.

Marginal lands previously used for agriculture can be planted to perennial plants; economic conditions and policy will determine whether these areas will be restored as prairie or planted as tame pastures. For example, subsidies offered through the CRP in the western United States or the Prairie Farm Rehabilitation Administration in Canada have resulted in the widespread planting of tame perennial grasses for soil conservation by paying farmers to plant and set aside marginal lands (Skold 1989). Planting tame grasslands has economic value as forage, and additional benefits of soil conservation and improved soil condition; however, we are also left with biotically impoverished grasslands dominated by a few exotic species. An alternative approach is to restore prairie.

Prairie Restoration

Attempts to restore prairie began in the 1930s at the University of Wisconsin's arboretum. Restoration usually involves some combination of reducing soil nutrient levels, obtaining propagules (i.e. seeds, sods) from local sources whenever possible, suppressing or eliminating weeds and exotic species, and maintaining an appropriate disturbance regime (e.g. fire, grazing and burrowing mammals) (Morgan et al. 1995, Wilson and Gerry 1995, Collins et al. 1998). Even with appropriate seed sources, sufficient labour, long-term economic support, and an appropriate disturbance regime, restoration may not recreate a native prairie. We need to compare the ecosystem functions of restored lands with native prairie, tame grasslands and agricultural lands to assess accurately both the long-term success of restoration efforts and to better understand changes in ecosystem functions from the losses of native prairie.

Prairie restoration may never replace native prairie entirely. For example, Kindscher and Tieszen (1998) found that plant species richness in Kansas was reduced by 3–6 species per m^2 in a 35-year-old restored area compared to adjacent native prairie; most of this decrease was caused by declines in nitrogen-fixing leguminous forbs. In addition, soil carbon and nitrogen levels were still much lower in restored areas than in native prairie. In a three-year restoration experiment in Colorado mixedgrass prairie, soil impoverishment and seeding of native species did not reduce the abundance of exotic plants (especially *Centaurea diffusa*, diffuse knapweed) (Reever-Morghan and Seastedt 1999). Of five native grasses seeded in this experiment, only one was present after three years (*Agropyron smithii*, western wheatgrass). Restoration efforts in many other ecosystems also show that even the best restoration attempts may only replace about half of the original species within a century (Bakker and Berendse 1999). A better measure of the success of prairie restoration would be to compare restored areas with successional prairie (i.e. areas once cultivated but allowed to undergo succession to native prairie species afterwards).

Prairie restorations do offer enormous opportunities for learning, adaptive management, and ecological research. Several important research questions need to be answered to evaluate the success of prairie restoration. First, is there a minimum area needed to establish a self-sustaining unit or community? Second, how closely does a restored system act like a native prairie with respect to ecosystem functions? Third, what disturbance regimes are necessary to maintain the species composition and functions of an ecosystem (see Ehrenfeld and Toth 1997)? While there are many unanswered questions surrounding the effectiveness of restoration, the effort is valuable as an educational tool and for generating public awareness of prairie.

Summary and Conclusions

Grasslands are formed regionally by environmental factors, but in the last century human management has had extensive impacts on the distribution of the northern native prairie. Grasslands provide a number of important ecosystem services which support continued environmental health and human livelihood. The losses of ecosystem services by replacing native prairie with tame grasslands dominated by exotic species or agricultural lands are poorly understood; however, if we recognize the importance of ecosystem services, we can explicitly value these critical ecological processes. The major points outlined in this article include the following:

1) Native prairie is one of the most highly modified and managed ecoregions in Canada, but is also an important source of biodiversity and ecosystem services to humans.
2) Conservation is easier than restoration, i.e. it is much easier to conserve native prairie than try to recreate or restore prairie. In addition, conservation of native prairie in several representative areas will provide critical baseline information against which we can measure the success of our management regimes.
3) Prairie conservation will require active management to maintain species composition, requiring the removal and elimination of exotic or weedy species, establishing buffer zones around native prairie wherever possible, and using appropriate combinations of fire and grazing.

4) Many of the ecosystem services provided by grasslands are belowground. There is little baseline data available on the ecosystem functions of grassland, but this information is critical to understanding changes in functions among different land management regimes or in restored prairie.

Many of these ideas have been previously suggested by land managers and grassland ecologists (Leopold 1949, Weaver 1954), and more recently in compilations by Joern and Keeler (1995) and Samson and Knopf (1996). Nearly fifty years ago, Weaver (1954: 325) warned that "the disappearance of a major unit of vegetation from the face of the earth is an event worthy of causing pause and consideration by any nation." Our best approach to prevent this from happening is through conservation and education, best viewed as "our effort to understand and preserve the capacity of the land for self-renewal" (Leopold 1949: 221). One powerful approach to conserving prairie is by recognizing, valuing, and better understanding the ecosystem services that it provides.

Acknowledgements

I wish to thank the Natural Resource's Institute at the University of Manitoba for logistical support, two anonymous reviewers and T. Radenbaugh for helpful comments on earlier drafts of the paper, and the University of Regina and the Natural Sciences and Engineering Research Council of Canada for financial support.

CHAPTER 5

HISTORICAL AND RECENT TRENDS IN THE AVIFAUNA OF SASKATCHEWAN'S PRAIRIE ECOZONE

Alan R. Smith and Todd A. Radenbaugh

ABSTRACT. Saskatchewan's prairies have a remarkably varied avifauna for a land-locked, high-latitude jurisdiction. This avifauna, however, has changed considerably over the last 150 years. Most of this change is due to direct and indirect effects of European settlement. Species whose populations have undergone marked changes are listed by trend patterns over long-term (1850-2000) and recent (1968-2000) periods. Where possible, the major factors that have influenced these trends are described.

Introduction

The Saskatchewan Prairie Ecozone is a temperate landlocked region lying near the centre of the North American continent. Although the avifauna landscapes here lack the diversity found in tropical or eastern North American forests, they do contain 246 bird species that regularly nest or visit on the way to the northern boreal forests (Smith 1996). A total of 184 prairie birds, most typical of the northern Great Plains, breed in the grassland, wetland, and woodland areas. There are, however, landscapes containing bird species more typical of other ecological regions. In the extreme southwest, species typical of the Great Basin Desert are encountered while along the Souris, Qu'Appelle and South Saskatchewan River valleys species typical of the eastern deciduous forest are historically found. One provincially unique region is the Cypress Hills, which contains many bird species characteristic of the Rocky Mountains.

As in many regions of the world, the avifauna of the Saskatchewan prairies changed as the landscape was altered by Europeans. Studies using the recent data (1965 to present) from the Breeding Bird Survey (BBS) have shown that populations of many endemic grassland birds have declined (Droege and Sauer 1994, Knopf 1994, 1995, Igl and Johnson 1997, Houston and Schmutz 1999). The causes of such bird population changes have not been directly measured, but habitat change on the breeding or wintering grounds and cyclical variations in populations have all been suggested.

In this article, we investigate nine of the pronounced trends observed in bird populations in Saskatchewan's Prairie Ecozone. These trends include extinct and extirpated species, population peak followed by decline, marginal species, long-term declines, decline with full recovery, decline with partial recovery, long-term increases, self-introduced species and introduced species. We also explore the possible reasons for these trends. Information is presented using two time references: long-term changes that extend back to the days of the earliest scientific expeditions in 1820s (expert opinion seen in Appendix A), and recent changes that followed the intensification of agriculture in the 1950s (Breeding Bird Surveys — BBS).

The Study Area and Its Avifauna

Saskatchewan's prairies are located on the northern extremity of the Great Plains of North America. The physical landscapes are primarily glacial in origin and include meltwater channels, lacustrine plains and moraines. Of particular note are the hummocky moraines, with their "knob and kettle topography" that form the "pothole" wetlands, one of continent's most important waterbird production areas. The climate is semiarid to humid continental, with long cold winters and short warm summers. The southern half of the province is part of the Prairie Ecozone and is dominated by prairie grasslands. The ecozone is divided into four ecoregions: the Cypress Upland, Mixed Grassland, Moist Mixed Grassland, and the Aspen Parkland. These areas are defined by similarities in climate, physiography, soil, and vegetation (ESWG 1995). Acton et al. (1998) described the abiotic and biotic features of each ecoregion while Radenbaugh (1998) described the major prairie plant assemblages. As one moves north-northeast, trembling aspen and other trees become more abundant and the grasslands gradually diminish over a 100 km ecotone (that includes the Aspen Parkland and Boreal Transition ecoregions).

Compared to other North American ecological regions, the northern Great Plains contain bird assemblages that are characterized by low species diversity, with only a few species having numerical dominance (Risser et al. 1981, Weins 1974, 1989, Cody 1985). This is partly due to the lack of abundant woody vegetation, which provides additional habitat dimensions and food sources. Wiens (1989), for example, found that within a region, by increasing the height and density of trees and shrubs both the richness and diversity of breeding birds will also increase. Historically, shrubby and treed areas were limited in the Saskatchewan prairies, with grasslands dominating between 80 percent to 95 percent of the landscape in the late 1800s (Archibold and Wilson 1980).

In the northern Great Plains, a large proportion (66 percent) of the breeding avifauna is comprised of either short- or long-distance migrants (MacArthur 1959, Wilson 1976). This is due to winters that are cold (resulting in higher energy demands) and low in food (due to low supplies or poor access due to snow cover) (MacArthur 1959). Moreover, unpredictable climatic "catastrophes" such as severe winters, early frosts, droughts and floods often reduce breeding bird populations (Wiens 1974). Such catastrophes also dampen bird populations in subsequent years, making summer food resources superabundant. Thus, summer food supplies are only rarely a factor that limits bird populations (Weins 1989).

Bird densities in the northern prairie systems are variable and respond to such perturbations as fire and grazing (Knopf 1996) and the introduction of exotic forages plants (Davis and Duncan 1999). The Mountain Plover and McCown's Longspur favour heavily grazed or disturbed grasslands, the Chestnut-collared Longspur prefers moderately grazed areas, while the Sprague's Pipit and Baird's Sparrow seek out taller grasses (Owens and Myres 1973, Sutter et al. 1995, Knopf 1996). Densities of breeding birds in the northern prairies have also been shown to

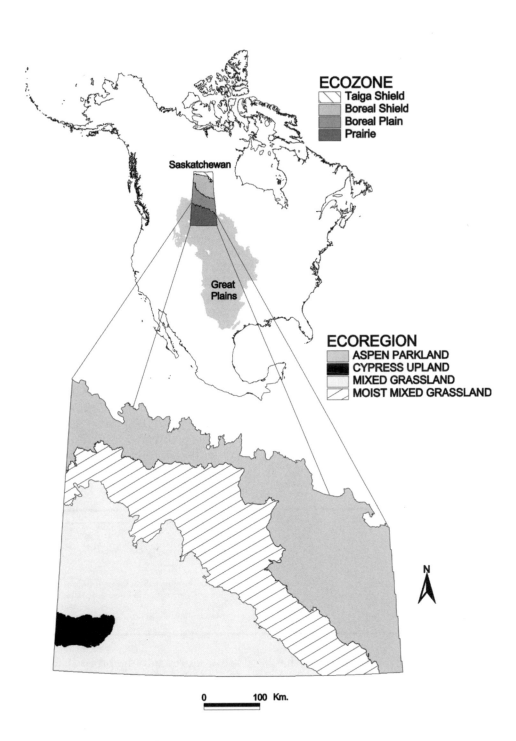

Saskatchewan's position within North America.

change with physiography. In North Dakota, for example, an average of 109 pairs of birds/km² nested on the relatively flat Agassiz Lake Plain, while an average of 272/km² used the hummocky moraines of the Missouri Coteau (Stewart and Kantrud 1972).

Long-term Changes in the Saskatchewan Avifauna (Mid-1800s to 2000)

In this article, descriptions of long-term changes are based on comparisons of data pre-dating or from the early years of European settlement to those of the present day. Appendix A lists the long-term trends for each species breeding in the ecozone. Data are based on a literature search and expert opinion. General data sources include the First (1819-22) and Second (1825-27) Franklin Expeditions (Richardson and Swainson 1831), and the observations of Coues (1878), Raine (1892), Macoun (1900), and Bent (1907, 1908). Other, more specific, sources follow the appropriate species accounts. As these sources contain almost no quantitative data, trends can only be inferred from subjective impressions of numbers along with gross estimates of habitat change. Fortunately, these anecdotes provide a sound basis for detecting the more marked changes such as the presence or absence of a species or a profound change in numbers. Similarly, early descriptions of habitat conditions along with historic vegetation cover data, derived from Archibold and Wilson (1980), can be used to infer trends for many species.

Extinct or Extirpated Species

Considering the profound influence that the arrival of Europeans had on the province's wildlife, it is surprising that only a handful of bird species has been lost (Table 1). The only species that has become extinct is the Passenger Pigeon, which bred in east-central Saskatchewan in the transition between the prairies and boreal forest (Houston 1972). More species have become extirpated from the region, include the Peregrine Falcon, Whooping Crane, and possibly the Mountain Plover.

Table 1. Extinct or extirpated bird species

Species	Extinct/extirpated	Nesting habitat	Cause(s) of decline
Peregrine Falcon	1917	Open country	Pesticides
Whooping Crane	1922	Wetlands	Loss of breeding habitat, overhunting, disturbance
Mountain Plover	Still extant?	Grassland (grazed)	Loss of breeding habitat through control of fires and elimination of bison and prairie dogs
Passenger Pigeon	1899	Woodland	Overhunting, loss of breeding habitat

Although the Peregrine Falcon has been reintroduced into the province in the cities of Saskatoon and Regina, it has not reoccupied its former range along Lodge Creek in the extreme southwest (Bechard 1981, 1982). The Whooping Crane was extirpated from its main breeding grounds on the northern Great Plains by 1922. However, a small number continue to breed in Wood Buffalo National Park, NWT, and often migrate through the region on their way to the southern coast of Texas. Although the Mountain Plover is now rare, the breeding status before European contact is debatable. This species prefers the short grass created by frequent fires, and heavy grazing by bison and prairie dogs (Graul and Webster 1976). Since those conditions were more prevalent 100 years ago, the species was probably more abundant than it is now.

Population Peak Followed by Decline or Extirpation

Table 2 lists three species that flourished during a temporary window between the extermination of the bison and the suppression of the frequent fires of the early settlement days (late 1890s to 1920s), and the intensification of agriculture that followed World War II. During this period, both grassland and woodland vegetation flourished interrupted only by the drought of the "Dirty Thirties." It was during this time that the Greater Prairie-Chicken and Dickcissel expanded into the landscape, but by the 1950s they had declined to become very rare as breeding species (Johnson and Smoliak 1976, Sealy 1971). Recently the Dickcissel has staged a small reinvasion in the southeast corner of the province (Smith unpublished data, 2000). The Whip-poor-will was, and still is, an uncommon breeding species in the Mid-Boreal Lowland Ecoregion; it once enjoyed a much wider range that extended south into the Aspen Parkland Ecoregion near Yorkton (Houston 1949).

Table 2. Bird populations that increased followed by decline or extirpation

Species	Arrival/Peak	Nesting habitat	Cause(s) of increase	Cause(s) of decline
Greater Prairie-Chicken	1895/1920s	Grassland	Fire suppression on breeding/ wintering grounds	Loss of breeding/ wintering habitat
Whip-poor-will	always present?/1930s	Woodland	Fire suppression on breeding grounds	Loss of breeding habitat
Dickcissel	1923/1930s	Grassland	Fire suppression on breeding grounds	Loss of breeding habitat

Marginal Species

Marginal species (Table 3) are those that maintain a temporary foothold in the region only to disappear when conditions become less favourable. Three of the five species are waterbirds, whose largest and nearest breeding concentrations are the marshes of the Great Basin (Price et al. 1995). Five years of severe drought (1985 to 1989) in the Great Basin seem to have forced them to search the northern Great Plains for suitable breeding sites. This speculation is supported by the fact that most of the Saskatchewan records of these species are from the mid to late 1980s (Smith 1996). Population fluctuations of the other two species, the Sage Thrasher and Common Poorwill, also appear to be related to drought cycles. Most of the records for these species (Kalcounis et al. 1991, Smith 1996) coincide with droughts in the 1930s, 1960s and 1980s.

Table 3. Ephemeral or marginal bird species

Species	Bred	Nesting habitat	Cause of temporary increase
White-faced Ibis	1976-78 (possibly), 1984-1988 (possibly)	Wetlands	Drought in Great Basin
Snowy Plover	1986-1989	Wetlands	Drought in Great Basin
Black-necked Stilt	1884, 1987, 1989, 1995	Wetlands	Drought in Great Basin
Common Poorwill	1964-65 (probably), 1983, 1982-1987 (probably)	Grassland	Droughts in the northern Great Plains
Sage Thrasher	1934, 1965, 1982	Sagebrush	Droughts in the northern Great Plains

Marked Long-term Declines

Species suffering long-term declines (Table 4) are primarily those affected by losses in the region and quality of breeding habitat due to agriculture. Grassland species have been particularly hard hit as the proportion of grassland in the Prairie Ecozone has been reduced to 20 percent of its original extent (Epp 1992, Selby and Santry 1996). Although some species such as the Horned Lark, and to a limited extent the McCown's Longspur and Western Meadowlark, have adapted to cropland habitats, many species have not. Recent and acute population downturns of the Sage Grouse and Burrowing Owl have been attributed both to loss and degradation of their grassland-breeding habitat (Braun 1998, Wellicome and Haug 1994). Sprague's Pipit also has been affected by losses in native vegetation since it is an area-sensitive species and requires large unfragmented native prairie (Davis et al. 1999). However, the decline in the Baird's Sparrow is more complex since the species occurs as frequently in hayland as in native and seeded pastures (Davis et al. 1999). Similarly, wetland drainage in the region has contributed to the decline of the American Bittern and to its principal prey species, leopard frogs (Smith 1996), while a combination of wetland drainage and cultivation of upland nesting cover has reduced numbers of Northern Pintail (Austin and Miller 1995). Population declines due to loss and fragmentation of nesting habitat are often exacerbated by increased nest predation. Cultivation has made it easy for predators such as the Red Fox, Raccoon and Striped Skunk (all themselves increasing in the region) to forage in the reduced, often linear, cover that agricultural wetlands provide (Beauchamp et al. 1996).

Marked Decline With Full Recovery

Species listed in Table 5 have declined due to human persecution, indiscriminate use of pesticides, and/or the extermination of the bison. Following major losses early in the 1900s some protection has been provided for these species. For example, policies that protect nesting colonies from human disturbance (e.g. commercial fishermen) have allowed the return of the American White Pelican. The enforcement of hunting regulations and releases of captive-raised birds have allowed for the rapid recovery of the Canada Goose. The banning of DDT and Dieldrin has allowed for the comeback the Richardson's Merlin in rural areas, while adaptation to new food sources and nest sites has allowed them to expand into urban areas (Houston and Schmidt 1981). The recovery of the Black-billed Magpie and Common Raven, on the other hand, is a testament to the adaptiveness and intelligence of these Corvids. The magpie had reoccupied its former range some 50 years ago (Houston 1977). The raven, for reasons unknown, is presently staging a comeback in the parklands and more wooded prairies and is expected to make a full recovery in the northern reaches in the near future.

Marked Decline With Partial Recovery

Table 6 lists bird species that declined in the first half of the twentieth century followed by a partial recovery starting in the 1970s. Causes of the declines are varied, while a common reason cited for the partial recovery has been the decline in Saskatchewan's farming population. Depopulation may have benefited large and shy birds such as Turkey Vulture, Trumpeter Swan and Sandhill Crane that require minimum disturbance during the breeding season. Although the Ferruginous

Hawk had been in decline from the 1950s to the 1980s (Houston and Bechard 1984), more recent data (Smith unpubl.) indicate that its numbers have partially recovered. However, this species of hawk is experiencing a long-term decline in reared young (Houston and Schmutz 1999) which may reverse this recovery. The mid-1990s decline of the introduced European Starling (see BBS results) may have allowed for some rebound in numbers of the Red-headed Woodpecker and Eastern Bluebird. As with the Richardson's Merlin (see previous section), the changing fortunes of the Loggerhead Shrike may be related to restrictions in the use of pesticides (Cadman 1986).

Table 4. Bird populations with marked long-term declines

Species	Nesting habitat	Cause(s) of decline
American Bittern	Wetlands	Loss of breeding habitat
Northern Pintail	Wetlands/Grassland	Loss of breeding habitat
Northern Harrier	Grassland/Wetlands	Loss of breeding habitat
Sharp-tailed Grouse	Grassland	Loss of breeding/wintering habitat
Sage Grouse	Grassland	Loss of breeding/wintering habitat
Upland Sandpiper	Grassland	Loss of breeding habitat
Long-billed Curlew	Grassland	Loss of breeding habitat
Burrowing Owl	Grassland	Loss of breeding habitat
Short-eared Owl	Grassland	Loss of breeding habitat
Sprague's Pipit	Grassland	Loss of breeding habitat
Baird's Sparrow	Grassland	Loss of breeding habitat
Chestnut-collared Longspur	Grassland	Loss of breeding habitat

Table 5. Bird populations with marked decline followed by full recovery

Species	Nesting habitat	Cause(s) of decline	Cause(s) of recovery
American White Pelican	Wetlands	Persecution	Protection
Canada Goose	Wetlands	Overhunting	Releases, protection
Richardson's Merlin	Grassland, urban	DDT, Dieldrin	Banning of pesticides; adaptation to urban habitats
Black-billed Magpie	Woodland, Grassland	Extirpation of bison	Adaptation to animal husbandry, urban habitats
Common Raven	Woodland, grassland	Extirpation of bison	Rural depopulation?

Table 6. Bird populations with marked decline with partial recovery

Species	Nesting habitat	Cause(s) of decline	Cause(s) of recovery
Turkey Vulture	Woodland, grassland	Extirpation of bison	Rural depopulation including abandonment of buildings
Trumpeter Swan	Wetlands	Overhunting	Protection
Ferruginous Hawk	Grassland	Human disturbance, loss of nesting habitat	Protection, rural depopulation, placement of nesting platforms
Sandhill Crane	Wetlands	Human disturbance	Rural depopulation
Red-headed Woodpecker	Woodlands	Competition with European Starling	Decline of Starling
Eastern Bluebird	Woodlands	Competition with European Starling	Decline of Starling
Loggerhead Shrike	Woodland, grassland	Loss of breeding habitat, pesticides?	Banning of pesticides?

Bird Populations With Marked Population Increases

Many species have undergone marked long-term population increases (Table 7). The reasons are varied. A proliferation of irrigation reservoirs in southwestern Saskatchewan has helped the Double-crested Cormorant as well as the American White Pelican. Eutrophication of wetlands may be responsible for increases in the numbers of Gadwall and Northern Shoveler (DuBowy 1996 and LeSchack et al. 1997). The increase of the Greater White-fronted, Snow and Ross' Geese seems to be due to increased food opportunities in the form of cultivated grains both on migration stopover areas and wintering grounds. However, the causes for the increase of the wetland-nesting Black-crowned Night-Heron remain obscure.

Although suppression of fires and planting of shelterbelts may have increased the populations of birds associated with woody cover, only a few have shown marked increases. The southward spread of groves of trembling aspen allowed for the expansion of the Red-tailed Hawk and Common Crow (Houston 1977, Houston and Bechard 1984). When the fire-protected landscape allowed trees to reach sufficient size for the development of nest cavities, the Great Crested Flycatcher (among other species) spread westward (Smith 1996). Further, the expansion of aspen groves and the planting of shelterbelts on a previously treeless landscape abetted birds such as the Mourning Dove, Western Kingbird and Lark Sparrow (Houston 1979, 1986, Houston and Houston 1979). Most of the swallow species have benefited from human activities such as road, bridge and building construction (see Plate 9) and the provision of nest boxes (Erskine 1979). The Violet-green Swallow might also be included as an increasing species, but it may simply have been undersampled in its remote badland habitat.

Self-introduced Bird Species

Several species (Table 8) have established themselves in the province since the arrival of Europeans. Some species, including the White-breasted Nuthatch and Yellow-throated Vireo have spread into the region because fire suppression changed the landscapes and created their breeding grounds. The Eastern Wood-Pewee seems to have increased because of fire control in Saskatchewan, as well as the clearing of wintering ground pasture from the tropical forests of Central and South America (Smith 1996). The Cattle Egret, a waterbird often associated with domestic ungulates, has been able to exploit a niche that was unfilled until its arrival in the Americas in the late 19th century.

Introduced Species

Ten species (Table 9) have been intentionally added to the landscape by humans. Five were introduced directly into the region, and the others arrived by expanding their ranges westward. The long-term viability of the Wild Turkey population remains in doubt, while the introductions of two other species, the Mute Swan and the Chukar, have been unsuccessful. The Eurasian Collared-Dove may become the eighth successfully established species. Most of these introduced species first showed a pattern of population increase followed by a slight decline.

Recent Trends — Breeding Bird Survey (BBS).

During the last 35 years, there have been a number of systematic bird surveys designed to detect population changes. Of these, the best and most comprehensive is the North American Breeding Bird Survey (BBS). Since its inception in 1966, the BBS has provided a useful tool for estimating abundance trends of breeding birds in North America. Routes are conducted only in areas that have allweather roads and two routes were randomly selected per "degree block" (one degree of longitude by one degree of latitude). Routes should be sampled annually during the

Table 7. Bird populations with marked population increase

Species	Nesting habitat	Cause(s) of increase
Double-crested Cormorant	Wetlands	Construction of reservoirs
Black-crowned Night-Heron	Wetlands	Unknown
Gadwall	Wetlands	Eutrophication of wetlands
Northern Shoveler	Wetlands	Eutrophication of wetlands
Red-tailed Hawk	Woodland	Spread of Aspen groves
Mourning Dove	Woodland/grassland	Planting of shelterbelts
Great Crested Flycatcher	Mature deciduous forest	Fire suppression
Western Kingbird	Woodland/grassland	Planting of shelterbelts
Purple Martin	Hollow trees/bird houses	Placement of bird houses
Tree Swallow	Hollow trees/bird houses	Placement of bird houses
Northern Rough-winged Swallow	Cutbanks	Building of roads (forming road cuts), excavation of gravel pits
Bank Swallow	Cutbanks	Building of roads (forming road cuts), excavation of gravel pits
Cliff Swallow	Cliffs, bridges, buildings	Construction of bridges and buildings
Barn Swallow	Cliffs, buildings, bridges	Construction of buildings and bridges
American Crow	Woodland/grassland	Spread of Aspen groves planting of shelterbelts
Lark Sparrow	Woodland/grassland	Planting of shelterbelts

Table 8. Self-introduced bird species

Species	1st record/1st nesting	Nesting Habitat	Cause(s) of increase
Cattle Egret	1974/1981	Wetlands/Grassland	Livestock husbandry
Eastern Wood-Pewee	1884/1960s	Mature deciduous forest	Habitat change on breeding and wintering grounds
Yellow-throated Vireo	1884/1990	Mature deciduous forest	Fire suppression on breeding grounds
Field Sparrow	1975/1980s	Brushy grassland	Fire suppression on breeding grounds
Orchard Oriole	1972/1974	Woodland/orchards	Planting of shelterbelts

Table 9. Exotic species introduced directly or indirectly into Saskatchewan

Species	Year of introduction or arrival*	Peak	Success	Nesting Habitat
Mute Swan	Introduced ca. 1956		Unsuccessful	Wetlands
Chukar	Introduced 1930s		Unsuccessful	Grasslands
Gray Partridge	Arrived 1921	1960s	Successful	Grasslands
Ring-necked Pheasant	Introduced 1920s	1960s	Successful	Shrublands
Wild Turkey	Introduced 1962	1980s	Successful?	Woodlands
Rock Dove	Introduced 1930s	1980s	Successful	Urban
Eurasian Collared Dove	Arrived 1998	Still increasing	Unknown	Urban
European Starling	Arrived 1937	1970s	Successful	Urban/Woodlands
House Finch	Arrived 1980	Still increasing	Successful	Urban
House Sparrow	Arrived 1899	1920s	Successful	Urban

* Year of direct introduction or arrival from introductions elsewhere.

height of the breeding season (usually in June) by skilled observers. Each route is 39 km (24.5 miles) long and consists of 50 stops spaced at 0.8 km (0.5 mile) intervals. Beginning 30 minutes before sunrise, counts are made at each stop for three minutes. During this time, all birds seen or heard within a 0.4 km (0.25 mile) radius of the stop are recorded.

The first BBS began in Saskatchewan in 1968 when 92 routes were selected. Of these, 72 had been conducted at least once as of 1999. Of 246 species of birds that regularly nest in Saskatchewan, over 210 have been recorded one or more times on BBS routes. Unfortunately, only 103 of these have been recorded in sufficient numbers and on enough routes to yield statistically significant trend data (Tables 10, 11, 12). The BBS provides insufficient data for the many species whose breeding ranges are restricted to poorly sampled regions, such as the Cypress Upland Ecoregion. Furthermore, the requirement for allweather roads biases results even within those regions where coverage is adequate. Since allweather roads tend to be located in heavily cultivated areas, surveys are biased for species frequenting farmland and edge habitats and biased against those that seek extensive grasslands, woodlands, badlands or marshes (for Saskatchewan see Sutter et al. 2000). Other limitations arise from the June date, causing the BBS to underestimate populations of some earlynesting birds, such as the Horned Lark. Further, as the BBS is primarily run in daylight hours, nocturnal birds like owls and goatsuckers are rarely observed.

In spite of its limitations, the BBS appears to have done a creditable job of monitoring populations of the more common bird species in heavily cultivated regions (e.g. southern Saskatchewan). Data from this survey show significant changes[1] in the populations of a number of species (Tables 10, 11). While the changes measured have not been even, they correlate with changes in habitat types. Birds have been classified as occurring in one of four habitat types based on nesting: Woodland-shrubland, grassland, wetland and other (cliffs, buildings or other artificial structures). If more than one habitat is used, the primary habitat (see Table 12) is listed first. Of the 12 species that have increased significantly from 1968 to 1996, seven nest in primarily woodland-shrubland habitats, three nest in wetlands, and only one nests in each of the other two habitat types (Tables 10 and 12). In contrast, only three of 17 significantly decreasing species use woodlands, the remaining are found in wetland and grassland habitats (Tables 11 and 12).

Although the species in all habitat categories show no significant population change it is difficult to *prove* that the measured bird trends are actually due to habitat change (Table 12). However, the BBS has only been conducted since 1968, and by this time agricultural influences on the landscape were large, with the majority of it being altered away from its native state. This, then, suggests that the major changes to bird populations had already occurred by the inception of the BBS. Therefore, one might speculate that since European settlement the area of woodland is increasing while that of wetlands and grasslands is decreasing, and the birds associated with these habitats may have responded accordingly. It follows that the few species with trends running counter to this might be responding to factors other than habitat change or caused by effects indirectly related to habitat change.

The strongest case that habitat change, specifically habitat loss, is affecting populations can be made for grassland birds. Almost one-third of the grassland species for which we have sufficient BBS data show steep declines. Several of the

Table 10. Species showing significant population increases (BBS data 1968-1996)

Species	Nesting habitat	Trend	P	Routes
Western Grebe	Wetland	5.4	n	4
Canada Goose	Wetland/grassland	15.9	*	23
Red-tailed Hawk	Woodland-shrubland	6.3	*	34
Cliff Swallow	Other	13.5	*	16
Mountain Bluebird	Woodland-shrubland	7.2	*	12
American Robin	Woodland-shrubland	5.6	*	41
Brown Thrasher	Woodland-shrubland	4.5	*	39
Spotted Towhee	Woodland-shrubland	13.6	*	10
Chipping Sparrow	Woodland-shrubland	10.7	n	23
Savannah Sparrow	Grassland/wetland	2.3	*	46
Yellow-headed Blackbird	Wetland	3.3	*	41
American Goldfinch	Woodland-shrubland	1.5	n	39

P = significance of trend, where n = p<0.05, * = p<0.01
Routes = number of routes used in calculating trend

Table 11. Species showing significant population declines (BBS data 1968-1996)

Species	Nesting habitat	Trend	P	Routes
Horned Grebe	Wetland	-5.4	n	24
Great Blue Heron	Wetland	-11.5	n	12
Northern Pintail	Wetland/grassland	-9.7	*	43
Northern Harrier	Wetland/grassland	-7.1	*	44
Sharp-tailed Grouse	Grassland	-6.5	n	21
Killdeer	Wetland/grassland	-2.8	n	45
Long-billed Curlew	Grassland	-7.3	*	6
Ring-billed Gull	Wetland	-3.2	n	31
Common Nighthawk	Grassland	-37.4	*	3
Belted Kingfisher	Wetland	-20.3	*	3
Northern Flicker	Woodland-shrubland	-7.1	*	30
Horned Lark	Grassland	-2.7	*	46
Sprague's Pipit	Grassland	-4.2	n	24
European Starling	Woodland-shrubland/other	-6.4	*	41
Lark Bunting	Grassland	-16.3	*	18
Song Sparrow	Woodland-shrubland	-3.5	*	35
Western Meadowlark	Grassland	-3.4	n	46

P = significance of trend, where n = p<0.05, * = p<0.01
Routes = number of routes used in calculating trend

Table 12. Summary of bird population trends by primary habitat type (BBS data 1968-1996)*

Trend	P	Woodland-shrubland	Wetlands	Grasslands	Other	Total
Increasing:	*	5	2	1	1	9
	n	2	1	0	0	3
Total Increasing		7	3	1	1	12
No Significant Change		25	30	14	5	74
Decreasing:	*	3	3	4	0	10
	n	0	4	3	0	7
Total Decreasing		3	7	7	0	17
Total Species		35	40	22	6	103

P = significance of trend, where * = p<0.01, n = p<0.05
* Includes only those species with sample sizes sufficiently large for analysis

decreasing wetland species also use upland nesting cover and their decline may be due as much to losses of grasslands as to wetlands. The good news for native woodland species is that numbers of the European Starling (an aggressive invader) are in decline. This should allow for increases in populations of the Red-headed Woodpecker and Eastern Bluebird as the introduced Starling competes for nest sites with the two native species.

Conclusions

Significant changes in Saskatchewan's avifauna correlate with the arrival of Europeans on the landscape less than 150 years ago. Most of these changes are likely due to European-style settlement and the agricultural development of the region. These activities have drastically altered the proportions of the major habitats in the region, as well as changed the quality of those habitats. Further, new habitats such as shelterbelts, urban areas, cultivated fields, and farmyards have been added, increasing the overall landscape heterogeneity.

Alteration in the proportion of habitats has generally translated into significant losses of non-agricultural and especially native habitats (i.e. grassland and wetland). More than 80 percent of the native grasslands have been ploughed and 50 percent of the wetlands have been drained for agricultural purposes (Epp 1992, Selby and Santry 1996). These losses have been offset somewhat by the planting of perennial forage crops (a grasslands substitute) and by the creation of numerous dugouts, and stock watering dams (wetlands substitutes). Although there has been significant clearing in the Aspen Parkland Ecoregion, the planting of shelterbelts and the natural expansion of native aspen groves has greatly increased the area of woodland to the south.

Alterations in habitat quality have had subtler effects, both positive and negative. The mostly negative effects included the conversion of prairie to croplands and the use of herbicides. Disturbance of nesting and roosting by humans and their pets coupled with competition from introduced species have also served to reduce the quality of the remaining native habitat. Habitat quality has been improved for species that use structures such as buildings and bridges, and by excavation for mines, gravel pits and garbage dumps. Land-use practices that have probably helped some species at the expense of others include the introduction of exotic plant species and changes in grazing, mowing and burning regimes. Further, many of the positive trends in species abundance have been the result of provincial, federal, and international programs targeting specific species, and more recently specific habitats. Traditionally these include the introduction of exotic and "game" species, feeding programs and the provision of artificial nest sites for song birds, and many have been designed to promote native species (e.g. Operation Burrowing Owl). Populations of rising "pest species" have been actively reduced through hunting, trapping and poisoning programs. Such measures may have to be entertained more frequently as burgeoning populations of Snow Geese and gulls become pests or threaten other populations or even entire habitats.

Habitat alteration by agricultural activity has become the most important engine of change on the landscape over the past 150 years. The proportion and number of habitats is presently different from those of the mid-eighteenth century. In light of this, one might ask if the functioning of this ecosystem been altered and if so, how have avian roles changed? Recent population declines (as shown by the BBS) of such common and adaptable species as the Killdeer and Western Meadowlark are cause for concern. Declines in these robust species, coupled with the increases in non-native populations, suggest that the very fabric of the prairie landscape is wearing thin and new assembly rules are developing. Further, since human alterations on the prairie landscape have affected the bird populations and their habitat directly, there could be cascading effects to other biotic groups and systems, including those upon which humans depend. Society has therefore, come to play an important ecological role in the prairie ecosystem that may be measured by changes and responses of component populations and their functional roles. Thus, society is a functioning component of the prairie landscape — and as one changes, so must the other.

Notes

1. The null hypothesis tested is that the trend does not differ from zero (no change). Rejection of the hypothesis indicates that there is evidence of a significant change in population size. An asterisk indicates that the chance that the trend is zero is less than one in 20 (*=p<0.05); an "n" less than one in 10 (n=p <0.10).

Appendix A

Breeding bird species of Saskatchewan's Prairie Ecozone, showing primary
nesting habitats, short-term (Breeding Bird Surveys)
and long-term (expert opinion) population trends

Common name	Primary Habitat	Short-term trend (BBS) 1968-96	P	Routes	Long-term trend 1819-2000	Long-term trend sources
Pied-billed Grebe	We	-2.1		27	↓	Macoun (1900), Mitchell (1923), Soper (1970), Roy (1996), Smith (1996)
Horned Grebe	We	-5.4	n	24	↔	Macoun (1900), Bent (1907), Mitchell (1923), Harrold (1933), Todd, 1947, Roy (1996), Smith (1996)
Red-necked Grebe	We	n/a			↔	Smith (1996)
Eared Grebe	We	6.5		15	↔	Macoun (1900), Bent (1907), Mitchell (1923), Todd (1947), Williams (1946), Roy (1996), Smith (1996),
Western/Clark's Grebe	We	5.4	*	4	↔	Raine (1892), Macoun (1900), Bent (1907), Mitchell (1923), Harrold 1933, Williams (1946), Nero *et al.* (1958), Roy (1996), Smith (1996),
American White Pelican	We	3.7		6	⇓⇑	Macoun (1900), Bent (1907), Mitchell (1923), Harrold (1933), Williams (1946), Roney (1993), Roy (1996), Smith (1996)
Double-crested Cormorant	We	-22.7		5	⇑	Macoun (1900), Bent (1907), Mitchell (1923), Roy (1996), Smith (1996)
American Bittern	We	1.0		28	⇓	Macoun (1900), Bent (1907), Mitchell (1923), Williams (1946), Roy (1996), Smith (1996)
Great Blue Heron	We	-11.5	n	12	↔	Macoun (1900), Bent (1907), Mitchell (1923), Harrold (1933), Williams (1946), Roy (1996), Smith (1996)
Cattle Egret	We/G	n/a			E	Hjertaas (1979), Roney (1982), Roy (1996), Smith (1996)
Black-crowned Night Heron	We	11.4		5	⇑	Macoun (19000), Mitchell (1923), Todd (1947), Roy (1996), Smith (1996)
White-faced Ibis	We	n/a			~	Smith (1996)
Turkey Vulture	Wo	n/a			⇓↑	Macoun (1900), Bent (1907), Mitchell (1923), Potter (1930), Roy (1996), Smith (1996)
Mute Swan	We	n/a			I/E	Smith (1996)
Trumpeter Swan	We	n/a			⇓↑	Smith (1996)
Canada Goose	We/G	15.9	*	23	⇓⇑	Macoun (1900), Bent (1907), Mitchell (1923), Harrold (1933), Williams (1946), Roy (1996), Smith (1996)
Wood Duck	Wo/We	n/a			↑	Smith (1996)
Gadwall	We	3.4		43	↑	Macoun (1900), Bent (1907), Mitchell (1923), Harrold (1933), Todd (1947), Roy (1996), Smith (1996) LeSchack *et al.* 1997
American Wigeon	We/G	-6.3		39	↔	Macoun (1900), Bent (1907), Mitchell (1923), Harrold (1933), Todd (1947), Williams (1946), Roy (1996), Smith (1996),
Mallard	We/G	-0.6		46	↓	Macoun (1900), Bent (1907), Mitchell (1923), Harrold (1933), Todd (1947), Roy (1996), Smith (1996)
Blue-winged Teal	We	0.4		42	↔	Macoun (1900), Bent (1907), Mitchell (1923), Harrold (1933), Todd (1947), Williams (1946), Roy (1996), Smith (1996)
Cinnamon Teal	We	n/a			↑	Macoun (1900), Bent (1907), Mitchell (1923), Roy (1996), Smith (1996)
Northern Shoveler	We	1.2		41	⇑	Macoun (1900), Bent (1907), Mitchell (1923), Harrold (1933), Todd (1947), Williams (1946), Roy (1996), Smith (1996) DuBowy 1996

Appendix A *continued*

Common name	Primary Habitat	Short-term trend (BBS)			Long-term trend	Long-term trend sources
		1968-96	P	Routes	1819-2000	
Northern Pintail	We/G	-9.7	*	43	⇓	Macoun (1900), Bent (1907), Mitchell (1923), Harrold (1933), Todd (1947), Williams (1946), Roy (1996), Austin and Miller (1995), Smith (1996)
Green-winged Teal	We	2.6		23	↔	Macoun (1900), Bent (1907), Mitchell (1923), Harrold (1933), Todd (1947), Roy (1996), Smith (1996)
Canvasback	We	0.0		29	↔	Macoun (1900), Bent (1907), Mitchell (1923), Harrold (1933), Roy (1996), Smith (1996)
Redhead	We	0.5		27	↔	Macoun (1900), Bent (1907), Mitchell (1923), Harrold (1933), Roy (1996), Smith (1996)
Ring-necked Duck	We	n/a			↔	Smith (1996)
Lesser Scaup	We	4.8		39	↔	Macoun (1900), Bent (1907), Mitchell (1923), Harrold (1933), Williams (1946), Roy (1996), Smith (1996)
White-winged Scoter	We	n/a			↓	Macoun (1900), Bent (1907), Mitchell (1923), Harrold (1933), Roy (1996), Smith (1996)
Common Goldeneye	We/Wo	n/a			↔	Smith (1996)
Bufflehead	W/Wo	n/a			↔	Smith (1996)
Ruddy Duck	We	2.3		26	↓	Macoun (1900), Bent (1907), Mitchell (1923), Harrold (1933), Williams (1946), Roy (1996), Smith (1996)
Northern Harrier	G/We	-7.1	*	44	⇓	Macoun (1900), Bent (1907), Mitchell (1923), Harrold (1933), Todd (1947), Roy (1996), Smith (1996)
Sharp-shinned Hawk	Wo	n/a			↑	Raine (1892), Macoun (1900), Bent (1907), Mitchell (1923), Harrold (1933), Williams (1946), Roy (1996), Smith (1996)
Broad-winged Hawk	Wo	n/a			↑	Smith. (1996)
Cooper's Hawk	Wo	-6.6		6	↑	Macoun (1900), Mitchell (1923), Roy (1996), Smith (1996)
Swainson's Hawk	G	0.8		41	↓	Macoun (1900), Bent (1907), Mitchell (1923), Potter (1930), Todd (1947), Williams (1946), Roy (1996), Smith (1996),
Red-tailed Hawk	Wo	6.3	*	34	⇑	Macoun (1900), Bent (1907), Mitchell (1923), Harrold (1933), Houston and Bechard (1983), Roy (1996), Smith (1996),
Ferruginous Hawk	G	-24.9		6	⇓↑	Macoun (1900), Bent (1907), Mitchell (1923), Potter (1930), Harrold (1933), Todd (1947), Houston and Bechard (1984), Roy (1996), Smith (1996),
Golden Eagle	G	n/a			↔	Macoun (1900), Bent (1907), Mitchell (1923), Potter (1930), Houston (1985), Roy (1996), Smith (1996)
American Kestrel	Wo/G	-6.1		14	↑	Macoun (1900), Bent (1907), Mitchell (1923), Harrold (1933), Roy (1996), Smith (1996),
Merlin	Wo/G	14.5		7	↓⇑	Macoun (1900), Bent (1907), Mitchell (1923), Potter (1930), Harrold (1933), Williams (1946), Huston and Schmidt (1981), Roy (1996), Smith (1996)
Prairie Falcon	G	n/a			↓	Macoun (1900), Bent (1907), Mitchell (1923), Potter (1930), Harrold (1933), Fyfe (1958), Roy (1996), Smith (1996)
Peregrine Falcon	G	n/a			X/I	Macoun (1900), Bent (1907), Williams (1946), Bechard (1981, 1982), Roy (1996), Smith (1996)
Chukar	G	n/a			I/E	Smith (1996)
Gray Partridge	G/Wo	6.5		30	I	Dexter (1922), Mitchell (1923), Potter (1930), Harrold (1933), Roy (1996), Smith (1996)

Appendix A *continued*

Common name	Primary Habitat	Short-term trend (BBS)			Long-term trend	Long-term trend sources
		1968-96	P	Routes	1819-2000	
Ring-necked Pheasant	Wo/G	-7.3		13	I	Roy (1996), Smith (1996)
Ruffed Grouse	Wo	n/a			↔	Smith (1996)
Sage Grouse	G	n/a			⇓	Macoun (1900), Bent (1907), Mitchell (1923), Roy (1996), Smith (1996)
Greater Prairie-Chicken	G	n/a			E,X	Macoun (1900), Mitchell (1923), Potter (1930), Harrold (1933), Todd (1947), Johnston and Smoliak (1976), Schmidt (1987), Roy (1996), Smith (1996)
Sharp-tailed Grouse	G	-6.5	n	21	⇓	Macoun (1900), Bent (1907), Mitchell (1923), Potter (1930), Harrold (1933), Todd (1947), Roy (1996), Smith (1996)
Yellow Rail	We	n/a			↔	Smith (1996)
Virginia Rail	We	n/a			↔	Smith (1996)
Sora	We	2.6		37	↓	Macoun (1900), Bent (1907), Mitchell (1923), Todd (1947), Roy (1996), Smith (1996)
American Coot	We	0.2		40	↔	Macoun (1900), Bent (1907), Mitchell (1923), Harrold (1933), Todd (1947), Williams (1946), Roy (1996), Smith (1996)
Sandhill Crane	We	n/a			⇓↑	Smith (1996)
Whooping Crane	We	n/a			X	Smith (1996)
Snowy Plover	We	n/a			~	Smith (1996)
Piping Plover	We	n/a			↔	Macoun (1900), Bent (1907), Mitchell (1923), Harrold (1933), Todd (1947), Williams (1946), Renaud (1979a), Skeel and Hjertaas (1993), Roy (1996), Smith (1996),
Killdeer	We/G	-2.8	n	45	↔	Macoun (1900), Bent (1907), Mitchell (1923), Harrold (1933), Todd (1947), Roy (1996), Smith (1996)
Black-necked Stilt	We	n/a			~	Smith (1996)
Mountain Plover	G	n/a			X	Graul and Webster (1976), Smith (1996)
American Avocet	We	6.1		21	↔	Macoun (1900), Bent (1907), Mitchell (1923), Harrold (1933), Todd (1947), Roy (1996), Smith (1996)
Willet	We/G	-1.3		44	↓	Macoun (1900), Bent (1907), Mitchell (1923), Harrold (1933), Todd (1947), Roy (1996), Smith (1996)
Spotted Sandpiper	We	-14.0		10	↔	Macoun (1900), Bent (1907), Mitchell (1923), Harrold (1933), Todd (1947), Roy (1996), Smith (1996)
Upland Sandpiper	G	-0.3		24	⇓	Macoun (1900), Bent (1907), Mitchell (1923), Harrold (1933), Todd (1947), Williams (1946), McNicholl (1988), Roy (1996), Smith (1996), Houston 1999
Long-billed Curlew	G	-7.3	*	6	⇓	Macoun (1900), Bent (1907), Mitchell (1923), Harrold (1933), Potter, 1930, Williams (1946), Todd (1947), Renaud (1980), Roy (1996), Smith (1996),
Marbled Godwit	G/We	-5.4		41	↓	Macoun (1900), Bent (1907), Mitchell (1923), Harrold (1933), Williams (1946), Todd (1947), Roy (1996), Smith (1996),
Common Snipe	We	-1.0		30	↔	Macoun (1900), Bent (1907), Mitchell (1923), Roy (1996), Smith (1996)
Wilson's Phalarope	We/G	-0.2		27	↓	Macoun (1900), Bent (1907), Mitchell (1923), Harrold (1933), Todd (1947), Roy (1996), Smith (1996)
Franklin's Gull	We	7.5		21	↔	Macoun (1900), Bent (1907), Mitchell (1923), Harrold (1933), Todd (1947), Roy (1996), Smith (1996)

Appendix A *continued*

Common name	Primary Habitat	Short-term trend (BBS)			Long-term trend	Long-term trend sources
		1968-96	P	Routes	1819-2000	
Ring-billed Gull	We	-3.2	n	31	↑	Macoun (1900), Bent (1907), Mitchell (1923), Harrold (1933), Todd (1947), Roy (1996), Smith (1996)
California Gull	We	12.4		10	↑	Macoun (1900), Bent (1907), Mitchell (1923), Harrold (1933), Todd (1947), Roy (1996), Smith (1996)
Common Tern	We	3.3		6	↓	Coues (1874), Macoun (1900), Bent (1907), Mitchell (1923), Harrold (1933), Todd (1947), Roy (1996), Smith (1996)
Forster's Tern	We	n/a			↑	Macoun (1900), Bent (1907), Mitchell (1923), Roy (1996), Smith (1996)
Black Tern	We	-3.7		30	↓	Coues (1874), Macoun (1900), Bent (1907), Mitchell (1923), Harrold (1933), Williams (1946), Todd (1947), Roy (1996), Smith (1996)
Rock Dove	O	-1.1		40	I	Bancroft (1993), Roy (1996), Smith (1996)
Eurasian Collared-Dove	Wo	n/a			I	Smith, A. unpublished data
Mourning Dove	Wo	1.5		45	⇑	Macoun (1900), Bent (1907), Mitchell (1923), Potter (1930), Harrold (1933), Todd (1947), Houston (1986), Bancroft (1993), Roy (1996), Smith (1996), Houston and Houston (1997)
Passenger Pigeon	Wo	n/a			X	Houston (1972), Smith (1996)
Black-billed Cuckoo	Wo	-2.1		20	↓	Coues (1874), Macoun (1900), Bent (1908), Mitchell (1923), Potter (1930), Roy (1996), Smith (1996)
Eastern Screech-Owl	Wo	n/a			↔	Smith (1996)
Great Horned Owl	Wo	-2.2		26	↑	Macoun (1900), Bent (1908), Mitchell (1923), Williams (1946), Houston (1996), Roy (1996), Smith (1996)
Burrowing Owl	G	n/a		0	⇓	Macoun (1900), Bent (1908), Mitchell (1923), Harrold (1933), Shaw (1944), Todd (1947), Wedgewood (1976), Houston et al. (1996), Roy (1996), Smith (1996)
Long-eared Owl	Wo	n/a			↓	Macoun (1900), Bent (1908), Mitchell (1923), Harrold (1933), Roy (1996), Smith (1996)
Short-eared Owl	G	-22.0		4	⇓	Macoun (1900), Bent (1908), Mitchell (1923), Todd (1947), Roy (1996), Smith (1996)
Northern Saw-whet Owl	Wo	n/a			↑	Macoun (1900), Mitchell (1923), Roy (1996), Smith (1996)
Common Nighthawk	G/O	-37.4	*	3	↓	Mitchell (1923), Williams (1946), Todd (1947), Wedgwood (1992), Roy (1996), Smith (1996),
Common Poorwill	G	n/a			~	Macoun (1900), Mitchell (1923), Smith (1996)
Whip-poor-will	Wo	n/a			E,X	Houston (1949), Smith (1996)
Ruby-throated Hummingbird	Wo	n/a			E	Macoun (1900), Mitchell (1923), Potter (1936), Roy (1996), Smith (1996)
Belted Kingfisher	We	-20.3	*	3	↔	Coues (1874), Macoun (1900), Bent (1907), Mitchell (1923), Williams (1946), Roy (1996), Smith (1996)
Yellow-bellied Sapsucker	Wo	n/a			↑	Smith (1996)
Downy Woodpecker	Wo	n/a			↑	Smith (1996)
Hairy Woodpecker	Wo	n/a			↑	Smith (1996)
Northern Flicker	Wo	-7.1	*	30	↑	Macoun (1900), Mitchell (1923), Todd (1947), Roy (1996), Smith (1996)
Red-headed Woodpecker100	Wo	n/a			⇓↑	Macoun (1900), Mitchell (1923), Williams (1946), Shadick (1980), Roy (1996), Smith (1996)

Appendix A *continued*

Common name	Primary Habitat	Short-term trend (BBS) 1968-96	P	Routes	Long-term trend 1819-2000	Long-term trend sources
Western Wood-Pewee	Wo	n/a			↑	Smith (1996)
Eastern Wood-Pewee	Wo	n/a			E	Houston (1980), Smith (1996)
Alder Flycatcher	Wo/We	n/a			↑	Smith (1996)
Willow Flycatcher	Wo/We	0.6		29	↑	Macoun (1900), Bent (1908), Mitchell (1923), Todd (1947), Roy (1996), Smith (1996)
Least Flycatcher	Wo	3.6		30	↑	Macoun (1900), Bent (1908), Mitchell (1923), Harrold (1933), Roy (1996), Smith (1996)
Eastern Phoebe	O	3.3		24	↑	Smith (1996)
Say's Phoebe	O	n/a			↑	Macoun (1900), Bent (1908), Mitchell (1923), Harrold (1933), Roy (1996), Smith (1996)
Great Crested Flycatcher	Wo	10.7		17	⇑	Smith (1996)
Western Kingbird	Wo	0.7		39	⇑	Coues (1874), Macoun (1900), Bent (1908), Mitchell (1923), Harrold (1933), Todd (1947), McKim (1926), Pearson (1936), Houston (1979), Roy (1996), Smith (1996)
Eastern Kingbird	Wo/G	-0.3		46	↑	Coues (1874), Macoun (1900), Bent (1908), Mitchell (1923), Harrold (1933), Williams (1946), Todd (1947), Roy (1996), Smith (1996),
Loggerhead Shrike	Wo	-2.6		30	⇓↑	Coues (1874), Macoun (1900), Mitchell (1923), Williams (1946), Todd (1947), Roy (1996), Smith (1996)
Yellow-throated Vireo	Wo	n/a			E	Smith (1996)
Warbling Vireo	Wo	1.3		26	↑	Coues (1874), Macoun (1900), Mitchell (1923), Harrold (1933), Todd (1947), Roy (1996), Smith (1996)
Red-eyed Vireo	Wo	-5.1		22	↑	Macoun (1900), Bent (1907), Mitchell (1923), Harrold (1933), Todd (1947), Roy (1996), Smith (1996)
Blue Jay	Wo	n/a			↑	Smith (1996)
Black-billed Magpie	Wo	0.8		41	⇓⇑	Swainson and Richardson (1831),Macoun (1900), Bent (1908), Henderson, (1923), Mitchell (1923), Farley (1925), Potter, (1930), Williams (1946), Todd (1947), Houston (1977), Roy (1996), Smith (1996)
American Crow	Wo	-3.0		46	⇑	Coues (1874), Macoun (1900), Bent (1908), Mitchell (1923), Potter, (1930), Harrold (1933), Williams (1946), Todd (1947), Houston (1977), Roy (1996), Smith (1996), Houston and Houston (1997)
Common Raven	Wo	n/a			⇓⇑	Swainson and Richardson (1831), Raine (1892), Macoun (1900), Potter (1930), Houston (1977), Roy (1996), Smith (1996)
Horned Lark	G	-2.7	*	46	↔	Macoun (1900), Bent (1908), Mitchell (1923), Potter, (1930), Harrold (1933), Williams (1946), Todd (1947), Roy (1996), Smith (1996),
Purple Martin	O/Wo	n/a			⇑	Smith (1996)
Tree Swallow	Wo	-1.2		33	⇑	Coues (1874), Macoun (1900), Bent (1908), Mitchell (1923), Harrold (1933), Houston and Houston (1997), Roy (1996), Smith (1996)
Violet-green Swallow	O	n/a			↔	Coues (1874), Smith (1996)
Northern Rough-winged Swallow	O	n/a			⇑	Macoun (1900), Mitchell (1923), Roy (1996), Smith (1996)

Appendix A *continued*

Common name	Primary Habitat	Short-term trend (BBS)			Long-term trend	Long-term trend sources
		1968-96	P	Routes	1819-2000	
Bank Swallow	O	-10.5		17	⇑	Coues (1874), Macoun (1900), Mitchell (1923), Potter, (1930), Harrold (1933), Todd (1947), Roy (1996), Smith (1996)
Cliff Swallow	O	13.5	*	16	⇑	Coues (1874), Macoun (1900), Bent (1908), Mitchell (1923), Potter, (1930), Roy (1996), Smith (1996)
Barn Swallow	O	-0.5		46	⇑	Coues (1874), Macoun (1900), Bent (1908), Mitchell (1923), Potter, (1930), Roy (1996), Smith (1996), Houston and Houston (1997)
Black -capped Chickadee	Wo	-2.7		10	↑	Coues (1874), Macoun (1900), Bent (1908), Mitchell (1923), Williams (1946), Todd (1947), Roy (1996), Smith (1996)
Red-breasted Nuthatch	Wo	n/a			↑	Smith (1996)
White-breasted Nuthatch	Wo	n/a			E	Smith (1996)
Rock Wren	O	n/a			↔	Macoun (1900), Mitchell (1923), Renaud (1979b), Roy (1996), Smith (1996),
House Wren	Wo	1.2		41	↑	Coues (1874), Macoun (1900), Bent (1908), Mitchell (1923), Harrold (1933), Williams (1946), Todd (1947), Roy (1996), Smith (1996)
Sedge Wren	We/G	n/a			↑	Smith (1996)
Marsh Wren	We	-2.1		6	E	Coues (1874), Macoun (1900), Mitchell (1923), Roy (1996), Smith (1996)
Eastern Bluebird	Wo	n/a			⇓↑	Smith (1996)
Mountain Bluebird	Wo	7.2	*	12	↑	Coues (1874), Macoun (1900), Mitchell (1923), Potter (1930), Harrold (1933), Williams (1946), Bittner (1988), Roy (1996), Smith (1996), Houston and Houston (1997)
Veery	Wo	10.8		11	↑	Coues (1874), Macoun (1900), Bent (1908), Mitchell (1923), Roy (1996), Smith (1996)
American Robin	Wo	5.6	*	41	↑	Coues (1874), Macoun (1900), Bent (1908), Mitchell (1923), Harrold (1933), Williams (1946), Todd (1947), Roy (1996), Smith (1996), Houston and Houston (1997)
Gray Catbird	Wo	1.5		23	↓	Coues (1874), Macoun (1900), Bent (1908), Mitchell (1923), Harrold (1933), Williams (1946), Roy (1996), Smith (1996)
Northern Mockingbird	Wo	n/a			E	Brazier (1964a) (1964b), Roy (1996), Smith (1996)
Sage Thrasher	Wo	n/a			~	Mitchell (1926), Roy (1996), Smith (1996)
Brown Thrasher	Wo	4.5	*	39	↑	Coues (1874), Macoun (1900), Bent (1908), Mitchell (1923), Harrold (1933), Roy (1996), Smith (1996)
European Starling	Wo/O	-6.4	*	41	I	Furniss (1944), Stewart (1951), Blue Jay (1957), Houston (1957), Roy (1996), Smith (1996)
Spraque's Pipit	G	-4.2	n	24	⇓	Coues (1874), Macoun (1900), Bent (1908), Mitchell (1923), Harrold (1933), Todd (1947), Roy (1996), Smith (1996), Davis et al. (1999)
Cedar Waxwing	Wo	1.9		21	↑	Coues (1874), Macoun (1900), Bent (1908), Mitchell (1923), Todd (1947), Roy (1996), Smith (1996)
Orange-crowned Warbler	Wo	n/a			↑	Smith (1996)
Yellow Warbler	Wo	1.0		36	↑	Coues (1874), Macoun (1900), Bent (1908), Mitchell (1923), Harrold (1933), Williams (1946), Todd (1947), Roy (1996), Smith (1996)

Appendix A *continued*

Common name	Primary Habitat	Short-term trend (BBS)			Long-term trend	Long-term trend sources
		1968-96	P	Routes	1819-2000	
Chestnut-sided Warbler	Wo	n/a			↑	Smith (1996)
Black-and-white Warbler	Wo	n/a			↑	Smith (1996)
American Redstart	Wo	n/a			↑	Smith (1996)
Ovenbird	Wo	-4.9		15	↑	Smith (1996)
Common Yellowthroat	We/Wo	-3.3		46	↑	Coues (1874), Macoun (1900), Bent (1908), Mitchell (1923), Harrold (1933), Williams (1946), Todd (1947), Roy (1996), Smith (1996)
Yellow-breasted Chat	Wo	n/a			↑	Mitchell (1923), Potter (1927), Williams (1946), Soper (1942), Todd (1947), Roy (1996), Smith (1996)
Rose-breasted/Black-headed Grosbeak	Wo	n/a			↑	Smith (1996)
Lazuli Bunting	Wo	n/a			↑	Mitchell (1924), Roy (1996), Smith (1996)
Indigo Bunting	Wo	n/a			↑	Smith (1996)
Dickcissel	G	n/a			E,X	Sealy (1971), Fretwell (1977), Smith (1996)
Spotted/Eastern Towhee	Wo	13.6	*	10	↑	Coues (1874), Macoun (1900), Mitchell (1923), Williams (1946), Roy (1996), Smith (1996)
Chipping Sparrow	Wo	10.7	n	23	↑	Coues (1874), Macoun (1900), Bent (1908), Mitchell (1923), Williams (1946), Todd (1947), Roy (1996), Smith (1996), Houston and Houston (1997),
Clay-colored Sparrow	G/Wo	1.0		44	↔	Macoun (1900), Bent (1908), Mitchell (1923), Harrold (1933), Todd (1947), Roy (1996), Smith (1996)
Brewer's Sparrow	G/Wo	n/a			↔	Mitchell (1923), Harrold (1933), Roy (1996), Smith (1996),
Field Sparrow	Wo/G	n/a			E	Stewart (1975), Skaar (1980), Smith (1996)
Vesper Sparrow	G	1.2		43	↔	Macoun (1900), Bent (1908), Mitchell (1923), Harrold (1933), Williams (1946), Todd (1947), Roy (1996), Smith (1996)
Lark Sparrow	Wo/G	-1.0		5	⇑	Macoun (1900), Houston and Houston 1979, Smith (1996), Knowles (1938)
Lark Bunting	G	-16.3	*	18	↔	Coues (1874), Macoun (1900), Bent (1908), Mitchell (1923), Williams (1946), Todd (1947), Nero (1993), Roy (1996), Smith (1996),
Savannah Sparrow	G/We	2.3	n	46	↔	Coues (1874), Macoun (1900), Bent (1908), Mitchell (1923), Harrold (1933), Todd (1947), Roy (1996), Smith (1996)
Baird's Sparrow	G	-7.0		33	↓	Coues (1874), Macoun (1900), Bent (1908), Mitchell (1923), Harrold (1933), Todd (1947), Davis et al. (1996; 1999), Roy (1996), Smith (1996)
Grasshopper Sparrow	G	-3.5		11	↔	Macoun (1900), Mitchell (1923), Harrold (1933), Todd (1947), Roy (1996), Smith (1996)
Le Conte's Sparrow	We/G	5.8		17	↔	Coues (1874), Macoun (1900), Mitchell (1923), Todd (1946), Roy (1996), Smith (1996)
Nelson' Sharp-tailed Sparrow	We	n/a			↔	Smith (1996)
Song Sparrow	Wo/We	-3.5	*	35	↔	Coues (1874), Macoun (1900), Bent (1908), Mitchell (1923), Todd (1947), Roy (1996), Smith (1996)

Appendix A *continued*

Common name	Primary Habitat	Short-term trend (BBS) 1968-96	P	Routes	Long-term trend 1819-2000	Long-term trend sources
McCown's Longspur	G	-10.5		9	↓	Coues (1874), Raine (1892), Macoun (1900), Bent (1908), Mitchell (1923), Harrold (1933), Todd (1947), Roy (1996), Smith (1996)
Chestnut-collared Longspur	G	-3.5		17	⇓	Coues (1874), Raine (1892), Macoun (1900), Bent (1908), Mitchell (1923), Harrold (1933), Roy (1996), Smith (1996), Davis et al. (1999)
Bobolink	G/We	3.0		29	↑	Coues (1874), Macoun (1900), Bent (1908), Mitchell (1923), Potter (1930), Roy (1996), Smith (1996)
Red-winged Blackbird	We	-0.4		46	↔	Coues (1874), Macoun (1900), Bent (1908), Mitchell (1923), Harrold (1933), Williams (1946), Todd (1947), Roy (1996), Smith (1996), Houston and Houston (1997)
Western Meadowlark	G	-3.4	*	46	↓	Coues (1874), Macoun (1900), Bent (1908), Mitchell (1923), Harrold (1933), Williams (1946), Todd (1947), Roy (1996), Smith (1996)
Yellow-headed Blackbird	We	3.3	*	41	↑	Coues (1874), Macoun (1900), Bent (1908), Mitchell (1923), Harrold (1933), Williams (1946), Todd (1947), Roy (1996), Smith (1996)
Brewer's Blackbird	G/Wo	-0.6		46	↓	Coues (1874), Macoun (1900), Bent (1908), Mitchell (1923), Harrold (1933), Williams (1946), Todd (1947), Roy (1996), Smith (1996)
Common Grackle	Wo	1.5		36	↑	Coues (1874), Macoun (1900), Bent (1908), Mitchell (1923), Harrold (1933), Todd (1947), Roy (1996), Smith (1996), Houston and Houston (1997)
Brown-headed Cowbird	Wo/G	-1.0		46	↑	Coues (1874), Macoun (1900), Bent (1908), Mitchell (1924), Harrold (1933), Williams (1946), Todd (1947), Roy (1996), Smith (1996)
Orchard Oriole	Wo	n/a			E	Coues (1874), Shadick (1980), Roy (1996), Smith (1996)
Baltimore/Bullock's Oriole	Wo	-0.1		40	↑	Coues (1874), Macoun (1900), Bent (1908), Mitchell (1924), Roy (1996), Smith (1996), Houston and Houston (1997)
Pine Siskin	Wo	n/a			↑	Smith (1996)
American Goldfinch	Wo	1.5	n	39	↑	Macoun (1900), Bent (1908), Mitchell (1924), Harrold (1933), Todd (1947), Roy (1996), Smith (1996)
House Sparrow	O	-0.3		45	I	Raine (1892), Macoun (1900), Mitchell (1924), Houston (1978, 1979), Roy (1996), Smith (1996)

Nesting Habitats: Wo= Woodland-shrubland, G = grassland, We = wetland, and O = other (cliffs, buildings or other artificial structures). If more than one habitat is used, the **primary habitat** is listed first.
Long term populations trends (circa 1819 to 2000): ↑ = increase, ⇑ = marked increase, ↔ = stable, ↓ = decrease; ⇓ = marked decrease, ⇓↑ marked decrease with partial recovery, ⇓⇑ marked decrease with full recovery, ~ = ephemeral or marginal, I = introduced, E = expanded range into Saskatchewan Prairies, and X = extirpated from Saskatchewan Prairies.
Short-term Trends (Breeding Bird Survey): P = significance of trend, where * = highly significant (p>0.01) change and n = significant (p>0.05) change; routes = number of routes used in calculating trend; n/a = data not available.
Note: Does not include species nesting only in the Cypress Hill Ecoregion or along the northern fringes of the Aspen Parkland Ecoregion are not included.

CHAPTER 6

THE CHANGING FACE OF PRAIRIE POLITICS:
POPULISM IN ALBERTA

Trevor Harrison

ABSTRACT. Populism has a long history in prairie politics. Recent years have seen the term resurrected in Alberta in the form of both the Reform party and Ralph Klein's Conservatives. This article compares the populism of early Social Credit in Alberta with the populism espoused by these recent political parties and their leaders. In doing so, the paper sheds light on how populist forms of discourse are both reproduced and transformed within a given political culture.

Introduction

When describing prairie politics historically, the term that readily comes to mind is "populism." For all its usage, however, populism remains an elusive concept. This is why a considerable body of literature over the years has tried to define different "types" of populism (for Canada, see Richards 1981; Finkel 1989; Laycock 1990; for the United States, see Canovan 1981).

Much less considered has been the manner in which populism has changed over time. An opportunity for chronological comparisons of populism is afforded by recent political events in Alberta. That province was the home of a "classic" instance of populism: the Social Credit party which governed Alberta between 1935–1971 (see Lipset 1950; Macpherson 1953; Mallory 1954; Irving 1959; Finkel 1989; Bell 1993). Recently, Alberta is the home of such populist-inspired parties as the Reform party (1987–2000) (Dobbin 1991; Manning 1992; Sharpe and Braid 1992; Patten 1993; Laycock 1994; Flanagan 1995; Harrison 1995a;) and Ralph Klein's Conservative government (1993–present) (Dabbs 1995; Laxer and Harrison 1995; Lisac 1995; and Kachur 1999). This article compares the populism of the early Social Credit (1935–43), ending with the death of its founder, the Reverend William Aberhart, with the populism of these recent political parties. In doing so, the article also suggests how populist forms of discourse are reproduced and transformed within a given political culture, concluding with speculation on the future of populism in Alberta.

Theorizing Populism

What is populism? A synthetic notion of populism embraces several key elements. All populist movements and parties involve a personal appeal by a leader to a mass audience. In some cases, populist leaders are "charismatic," but this is not necessarily the case. Central to the leader's appeal is the notion of "the people," a group defined by its historic, geographic, and/or cultural roots. This appeal is made urgent by the perception of a crisis threatening "the people." Finally, the source of this threat is another group — sometimes termed "power bloc" — viewed as physically or culturally external to "the people." Typically, from a left-wing perspective, the threat is posed by capitalism or "big business"; and from a right-wing perspective, by "big government" (Laclau 1977; Conway 1978; Sinclair 1979; and Patten 1993). Populism is thus defined as: 1) a mass political movement 2) based on an imagined personal (i.e., unmediated) relationship between leaders and followers 3) mobilized around symbols and traditions congruent with the popular culture 4) which expresses a group's sense of threat 5) arising from powerful external elements.

Despite the apparent precision of this definition, populism in practice tends to be quite diverse. Leadership ideologies and styles vary, from Benito Mussolini to Juan Peron to George Wallace. Definitions of both "the people" and the opposing "power bloc," and the precise nature of the impending threat, are similarly elastic, being constructed in the course of real political struggles. The characteristics of supporters (occupation, class, etc.) are likewise important determinants of the kind of populist party that emerges. Finally, parties and movements also have a "career" (Clark et al. 1975; Manning 1992: 50–51). As with individuals, both success and failure can alter career trajectories. Populist parties may become bureaucratized and routinized; in Zakuta's (1975) words, their populist impetus may be "becalmed." Taking these considerations into account, several attempts have thus been made to classify "types" of populism.

Typologizing Prairie Populism

Canadian typologies of populism have been frequently constructed along traditional Right-Left lines. Seminal in this regard is Richards' (1981) characterization of populist movements. Specifically, Richards notes that, historically, Left populism in Canada has sprung from rural co-operative organizations; involved class (farm-labour) alliances; and critiqued corporate capitalism, demanding a greater role for the government and state to counter corporate power. In contrast, Right populism has tended to mobilize along regional rather than class lines. Right populism restricts its corporate critique to the power of banks, the money supply, and credit, and views big government as the primary enemy of the people. Unlike left populism, right populism downplays participatory democracy in favour of measures of direct democracy, such as plebiscites, referendums, and electoral recall.

Finkel (1989) subsequently applied Richards' (1981) right-left typology of populism to his examination of the Alberta Social Credit party. Laycock (1990) later expanded Richards' (1981) typology into four categories — social democratic populism, radical democratic populism, plebiscitarian populism, and crypto-Liberalism. Table 1 provides a modified summary of Richards' (1981) and Laycock's (1990) populist types.

Typologies of prairie populism are, almost invariably, based upon populist experiences drawn from the period between World War I and the end of World War II. Alberta is a thread connecting specifically many of these experiences.

Alberta populism was influenced by American immigrants in the early part of

Table 1. A Modified Version of Richards's and Laycock's Populist Typologies

	Richards's (1981) Typology			Laycock's (1990) Typology		
	Left	Right	Social Democratic	Radical	Crypto-Liberal	Plebiscitarian
Basis of Electoral Alliance	class: farm-labour	region: west (vs. east)	class: farm-labour	farm	region: rural-hinterland vs. central Canada	region: west-east
"The People"	non- (big) capitalists	non-elites	non- (big) capitalists	non-elites	agricultural producers	consumers
"The Power Bloc" (Political Critique)	general: corporate capitalism	specific: banks, big government	general: corporate capitalism	general: gov't., capitalism	limited: eastern business	specific: federal government, banks
Chief Reforms Advocated	break up trusts, government intervention	increase credit democracy ensure competitive markets	Fabian socialism	group, delegate	free trade, minimal gov't	increase credit
Organizational Form/ Mechanisms for Political Participation	co-operatives	direct democracy: plebiscites, recall referendums	farm organisations, participatory democracy	rural communities co-operatives unions	farm organisations occupations	rule by experts
Example	Cooperative Commonwealth Federation (1932-44)	Alberta Social Credit (1935-43)	Cooperative Commonwealth Federation (1932-44)	Alberta Non-Partisan League (1917-1919)	National Progressive party (1921-1930)	Alberta Social Credit (1935-43)

the twentieth century (Conway 1978), several of whom were veterans of the original United States populist movements. This made Alberta a hotbed of radical political experimentation almost from its creation in 1905. In the provincial election of 1917, the Non-Partisan League elected two candidates on a platform that joined socialist policies of intervention in the economy with a belief in non-party politics.

Two years later, the League merged with the United Farmers of Alberta (UFA) (Bell 1993: 11) which, taking a formally political stand, captured office in 1921. The UFA was an offshoot of wider agrarian unrest that saw farmers' parties also elected in Ontario (1919) and Manitoba (1922), and led to the National Progressive party garnering 65 seats in the 1921 federal election. From the beginning, however, the Alberta contingent was more politically radical than its counterparts elsewhere in Canada, espousing notions of group government and a mistrust of the entire party system (see Morton, 1950).

During the Depression of the 1930s, Alberta gave birth to two more populist movements. In 1932, a meeting in Calgary provided the impetus for the formation of the Cooperative Commonwealth Federation a year later (CCF) (Lipset 1968; Laycock 1990; also Richards and Pratt 1979).[1] Then, in 1935, Social Credit seized power in Alberta, throwing out the UFA and beginning a reign that did not end until 1971 (Macpherson 1953; Finkel 1989; Bell 1993).

It is argued that populism remains a fundamental feature of prairie political culture (Gibbins 1980; Laycock, 1990; Gibbins and Arrison 1995). Yet, from the 1950s to fairly recently, parties formally espousing populism on the prairies (and elsewhere) were rare. The trend back to populism began in the United States with the civil rights movements of the 1960s and the Reagan-led, neo-conservative countermovements of the late 1970s. A while later, the term "populism" also was resurrected in Canada. As before, this populist revival took root in Alberta, first with the formation of Preston Manning's Reform party in 1987 and, later, with the election of Ralph Klein's Conservatives in 1993.

The emergence of these parties provides the basis for considering both how populism has changed over time and how populist forms of discourse are reproduced within a political culture. Specifically, this paper compares the populism of early Social Credit with the forms of populism espoused and practiced by the federal Reform party and the Alberta Conservative government of Ralph Klein.

United Farmers and Equity Association of Alberta.

The three parties are compared according to four criteria: 1) style of leadership; 2) characterizations of "the people"; 3) characterizations of the "power bloc" and the threat posed; and 4) the form of organization and mechanisms for political participation underlying the parties.

Leadership Style

Given the fact that his father (the Reverend Ernest Manning) succeeded the Reverend William Aberhart as "Socred" premier, one might assume Preston Manning's leadership style to be similar to that of Aberhart. This comparison is correct to a point. Aberhart was a spellbinding orator; likewise, Preston Manning — if not captivating - is nonetheless a polished speaker able to draw upon historical references and symbols, wrapped in a kind of folksy humour. By contrast, Klein is a master of the sound bite and the *ad hominem*, but in formal circumstances is a rather pedestrian speaker.

Also, like Aberhart, Preston Manning is devoutly religious (for Aberhart, see Irving 1959; Finkel 1989; for Manning, see Dobbin 1991), though he lives his faith fairly privately. By contrast, Ralph Klein's demeanor and approach are decidedly secular (see Dabbs 1995; Lisac 1995). Still, one should note that some of Klein's chief lieutenant's (e.g., Stockwell Day, former Alberta Treasurer and now leader of the new Canadian Alliance Party), wear their fundamentalist religious beliefs on their sleeves.

Granting these observations, Klein and Aberhart are also similar. For example, Aberhart was not a "details" man. He quite avowedly relied upon experts to implement social credit policy (Finkel 1989; Bell 1993). Klein similarly relies upon a small group of advisers and trusted confidants, and seems otherwise bored with the minutia of government (Dabbs 1995; Lisac 1995). By contrast, Preston Manning's style of leadership is very hands-on and technical. Indeed, he is widely viewed to be a "micro-manager," poor at delegating and tending instead to centralize control

Provincial Archives of Alberta P5313
Alberta Premier William ("Bible Bill") Aberhart, 1935.

(Harrison 1995a; Flanagan 1995). In part, Manning's greater need than Aberhart or Klein for control appears related to his belief that a populist party is open to seizure by extremist elements (Manning, 1992: 260; also, see below).

Klein more than Manning also shares with Aberhart "the common touch." Klein's concern for the "ordinary person" seems genuine, and comes across whether in small, face-to-face groups or on television. For many Albertans, Klein is just one of them (Dabbs 1995; Lisac 1995). By contrast, Manning often seems awkward, even shy, around people (see Harrison 1995a); yet, as with Klein and Aberhart, Manning's followers are personally and devoutly loyal to him.

An ability to connect personally with their followers is reflected in the three leaders' respective use of the media. In 1925, Aberhart was already a well-known fundamentalist preacher. That year, he began hosting a weekly radio broadcast from Calgary. Within a short time, the "Back to the Bible Hour" was being heard by more than 300,000 people, 65 percent of whom lived in Alberta (Finkel 1989: 29). Freely mixing politics with Biblical injunctions, Aberhart later used his radio program to bypass the mainstream media — most of which was extremely hostile to social credit doctrine (Conway 1994: 121)[2] — to "sell" his ideas. After the 1935 election, Aberhart continued his weekly religious broadcasts, a tradition taken up by Ernest Manning when he succeeded Aberhart as premier in 1943 (Finkel 1989; Bell 1993).

Like Aberhart, Klein also had public exposure before formally entering politics. From the late 1960s until 1980, he was a television news reporter in Calgary, a job which gave him not only name recognition but also practical knowledge of how to use the media. In Calgary's 1980 civic election, he rode the public dissatisfaction with his opponents and his own public profile as a man of the people – a "straight shooter" — to the mayor's chair. After successful re-elections in 1983 and 1986, Klein jumped to provincial politics in 1989, being elected Conservative MLA. In December 1992, Klein succeeded Don Getty as Tory leader and premier of the province; in 1993, he led the Conservatives to victory (see Dabbs 1995; Stewart 1995), a feat he repeated in 1997.

Like Aberhart, Klein frequently employs the media to speak directly to the people. Once a year, he makes a televised address to the people of Alberta — always on a private channel which is under no regulatory obligation to offer equal time to political opponents. Bracketing this singular event are "selective" television broadcasts stating the Conservative party views on issues of importance, or occasional radio interviews with hosts known to be supportive of the Tories. Finally, in a particularly secular version of the Back to the Bible Hour, Klein also hosts a weekly

Provincial Archives of Alberta 67.113
A "Prosperity Certificate" from 1936 ("Prosperity Certificates" were a Social Credit innovation designed to counteract the influence of the big banks).

radio show ("Talk to the Premier") which is broadcast on a private station throughout the province.

Where Aberhart feuded with big business and the mass media, however, Ralph Klein's Conservative party is the darling of big business and the corporate media. During Klein's first two years in office, the *Wall Street Journal*, Barron's, the *New York Times*, and the *Globe and Mail* regularly published glowing accounts of the "Klein revolution" (see Harrison and Laxer 1995: 2). Within Alberta, the Calgary Herald and respective Sun newspapers in Calgary and Edmonton have been consistently pro-Klein and pro-Conservative. Among major newspapers, only the *Edmonton Journal* has voiced opposition to the government.

During Reform's 13-year existence, Preston Manning never garnered regular coverage from the national and pro-business media. When he did receive coverage, it was usually negative or sensational. Reform lacked real political power, even while in the office of Official Opposition, and thus was not newsworthy. The party also was unable to shake its image as a Western fringe party. In this context, Reform received consistently positive media coverage only from the *Alberta Report* (and its affiliates)[3] and a number of rabid talk show hosts, mainly in Western Canada.

Like Aberhart, Preston Manning tried to overcome the media gap by using the latest "cutting edge" technologies. For example, Reform was one of the first political parties in Canada to create its own website; it also pioneered attempts at televised nation-wide town hall meetings (see Barney 1995). These efforts at manufacturing a personal relationship between Preston Manning and the masses generally failed. Nonetheless, in a curious way, Manning's failure to appeal to *all* Canadians reinforced in the eyes of Reform's "true believers" the unassailable correctness of their beliefs. Prophets, after all, are seldom recognized in their own time.

The People

Social Credit implied two specific characterizations of "the people": "consumers" (Laycock 1990) and "non-elites." The first characterization of the people as consumers derived from Social Credit theory's assertion that the monetary supply needed to be expanded in order to fuel consumer demand. The second characterization of the people as "non-elites" expressed a traditional sense of powerlessness, especially in relation to "big business" (banks, railroads, etc.) and "big government." The west in 1935 was more a region, and less given to provincialism, than it would be only a few decades later (Gibbins 1980). Though Social Credit's policies were largely contained within Alberta for electoral reasons, initially its message was broadly pitched at non-elites everywhere.

Like Social Credit, Reform from its inception also implied two main characterizations of the people, "westerners" and "non-elites" (i.e., "ordinary" Canadians). Reform's characterization of westerners as "the people" reflected traditional regional alienation, and was enunciated in the party's founding slogan in 1987: "The West Wants In" (Sharpe and Braid 1992). Reform advanced the notion of representing the entire West, despite the fact that its support was disproportionately located in the two "have provinces" of Alberta and, to a lesser degree, British Columbia (Harrison 1995a).

After 1991, when Reform voted to expand into a national party, the characterization of the people as "westerners" was dropped in favour of a broad appeal to "non-elites." Unresponsive politicians, appointed Senators, and assorted "special interests" (Harrison et al. 1996) were blamed for Canada's economic and political woes. Taxes were a chief source of complaint, as elites were widely denounced for wasting the hard-earned money of ordinary, middle-class Canadians. Reform's first electoral victories in 1989 — a by-election win (Deborah Grey) in Beaver River, Alberta and an election (Stan Waters) to fill a vacant Alberta seat in the Senate — were ignited by opposition to the Goods and Services Tax (GST) (Harrison 1995a). Reform similarly garnered support in 1990 and 1992 for opposing the Meech Lake and the Charlottetown Accords, which it argued were being imposed by an arrogant political elite (Harrison 1995a; Flanagan 1995).

Like Social Credit and Reform, the Klein government also presents two characterizations of the people: "stakeholders," and "ordinary"[4] Albertans. In broad terms, stakeholders are "builders," "entrepreneurs," and "boosters." In concrete terms, the stakeholder characterization includes taxpayers, both individual and corporate, which the Klein government argues must be rewarded for their hard work through a regime of low taxes[5] (Laxer and Harrison 1995). But the stakeholder characterization of the people also gives pride of place to private sector decision-makers and goods/service providers (i.e., corporations, lobbyists, and associations). In this vein, the Klein government since 1993 has frequently called together gatherings of stakeholders at various "summits," "roundtables," etc. to discuss policy directions for the province (see Kachur 1999). These meetings involve a subset of the province's intellectual, corporate, and government elite. The opening of the legislature in the fall of 1997 was foregone in favour of one such meeting, "the Growth Summit."

Alberta's unelected elites regularly meet to design government policy. Given this fact, the Klein government cannot characterize the people as "non-elites." At the same time, the notion that everyone is just an "ordinary" Albertan acts to level — more properly, ignore — existing class, gender, ethnic, or other differences.

The belief that no fundamental social differences, and therefore no real bases for conflict, exist runs deep in Alberta. This belief can be found, for example, in

Preston Manning's approach to politics, an approach which political scientist Tom Flanagan (1995) refers to as Manning's notion of "monism," or oneness. According to Flanagan, Manning does not recognize the existence of essential fractures within the voting public (see also Dobbin 1991).

This "monistic" belief in essential harmony, and a refusal to acknowledge differences and conflict, has had real historical consequences in Alberta. While it has been useful in mobilizing Albertans against outside threats, this belief also has resulted in a long history of one-party dominance. Only four parties historically have governed Alberta, each for long periods of time[6]; and unlike other provinces, Alberta has never experienced a minority government. This "monistic" belief also has undermined the notion of political parties themselves and of the party system; indeed, political parties in Alberta are dismissed as *causes*, rather than *expressions*, of social and political division. In the early days, for example, the Non-Partisan League and United Farmers of Alberta both ran on platforms of anti-party politics (Finkel 1989: 19). Social Credit followed in what Macpherson (1953) termed Alberta's "quasi-party" system. After the Socreds fell in 1971, the Tories under Peter Lougheed (1971–84) and Don Getty (1984–92) continued to nurture the idea that the government caucus was sufficiently diverse to represent the views of all Albertans. This notion has been repeated in recent years by Ralph Klein who has openly mused that Opposition members of the legislature are unnecessary (see Epp 2000). At the same time, however, definitions of "us" and "them" in Alberta since 1993 have taken an interesting turn: the pretense of one party representing all Albertans has been severely tested, as we will see.

The Power Bloc and the Threat

The creation of the Social Credit in 1935, the formation of the Reform party in 1987, and the reconstruction of the Tories under Ralph Klein in 1993 arose during periods of political and/or economic upheaval. In the case of Social Credit, Alberta and Canada in 1935 were still in the grip of the Great Depression. On the prairies, the Depression was made worse by years of severe drought; workers were faced with high rates of unemployment (20–25 percent); farmers and small businesspeople faced foreclosures and bankruptcy. All felt threatened by an economic system that seemed increasingly alien. Yet, Canada's political leaders — R.B. Bennett in Ottawa, John Brownley in Alberta — seemed unresponsive to their plight (see Berton 1991; Conway 1994). In a context of growing fear and unrest, Social Credit provided an explanation for the economic crisis: a lack of consumer spending power. Social Credit also identified the power bloc benefitting from the current crisis and thereby threatening the people: first, banks and bankers; second, "big" government.

Two points must be made regarding the place of banks in Socred philosophy. First, while some of Alberta Social Credit's followers held socialist ideas (Finkel, 1989), its formal doctrines (as enunciated by Major Douglas) were largely reformist. Neither capitalism nor "big business" was at fault *per se*; rather, the problem lay with a few actors in the system. The system itself simply needed fine-tuning. Thus, while Aberhart denounced "poverty in the midst of plenty," he specifically condemned "the Fifty Big Shots of Canada" — bankers and financiers — who profitted by high interest rates.

The second point is that Social Credit's focus upon banks, and more personally upon bankers and financiers, merged easily with culturally entrenched beliefs of a "Jewish conspiracy." Major Douglas' own writings were blatantly anti-Semitic. So, too, were some statements made by prominant Social Credit MLAs in Alberta after 1935 (Irving 1959; Finkel 1989; Bell 1993; Stingel 2000). In 1948, Ernest Manning publicly denounced anti-Semitism within the party and purged the party of several of its most extreme members (Finkel 1989: 104–6). Nonetheless, it is clear that, for many Social Crediters, banks were synonymous with Jews as the "power bloc" threatening the people.

That "big government," especially the federal government, was also viewed as part of the power bloc is not surprising. After all, it was held that the west was founded (under the National Policy of 1879) as an economic colony of central Canada. Indeed, ownership of its natural resources had only recently (1930) been transferred to Alberta by the federal government. This antipathy, however, towards government and, more broadly, towards the state was reinforced after Social Credit came to power.

In 1936, Social Credit defaulted on paying the holders of some provincial bonds, and followed this by unilaterally cutting the interest rate on Alberta's public debt (Conway 1994). These experiments in populist radicalism were only the beginning, however: faced with growing unrest from a group of insurgent MLAs anxious to see Social Credit ideas implemented, Aberhart's government in 1937 passed a series of laws, many of which challenged the rights of private property and big business, including the media[7], as well as federal jurisdiction.

Conway (1994: 122) summarizes thus the results of this period of radicalism:

> Eleven provincial statutes were disallowed by the federal Cabinet, while the Alberta Lieutenant-Governor refused to sign three duly passed bills into law. Countless others were taken to court in a series of legal battles, all of which the Social Credit government lost. The whole package of Social Credit laws was finally declared *ultra vires* by the Supreme Court in 1938. Three years later the Court struck down Aberhart's elaborate debt adjustment legislative edifice.

Little wonder then that supporters of Social Credit viewed their efforts at radical reform as being thwarted by the combined efforts of the financiers and their federal government collaborators.

Social Credit's depiction of the opposing power bloc was not constructed out of whole cloth. Western hatred of banks, railroads, eastern manufacturers, and Ottawa had deep roots. Likewise, while the Reform party arose in 1987, the party's ideological and political roots went back much further.

In the 1962 Canadian election, Lester Pearson's Liberals defeated John Diefenbaker's Tories. Thereafter, the Liberal party governed Canada almost uninterrupted for the next 22 years. Yet, during this time, the West — especially Alberta — remained solidly Tory. People in Alberta believed that one day "their party" would regain office and redress the perceived wrongs. Thus, many Albertans "endured" the long winter (as they saw it) of Liberal rule, including especially Pierre Trudeau's introduction in the early 1980s of the hated National Energy Program.

In 1984, the Tories regained office under Brian Mulroney. Many conservative supporters in Alberta believed that policies they favoured, such as free trade, a more decentralized federation, smaller government, an end to bilingualism and multiculturalism, and a tougher stand on Quebec separatism would now be implemented. And, indeed, some of these policies (e.g., NAFTA) were enacted. But discontent continued, reaching new heights with the signing of the Meech Lake Accord in 1987 and the passing of the Goods and Services Tax two years later. In the eyes of many fiscal and moral conservatives, the Tories were merely Liberals in disguise. Thus, Reform was launched in 1987 on twin platforms of both regional discontent and an explicit promise to resurrect a "genuinely" conservative Canada (see Dobbin 1991; Sharpe and Braid 1992; Harrison 1995a).

Although Reform failed to win a seat in the 1988 Canadian election, the party continued to garner support. The 1993 election saw Reform capture 52 seats, finishing third to Jean Chretien's Liberal party and the opposition Bloc Quebecois. Then, in 1997, Reform took 60 seats, becoming Canada's Official Opposition. It was a remarkable feat for a party formed only ten years earlier.

Underpinning Reform's rise to political prominence was the party's articulation of the power bloc threatening Canada and Canadians. This power bloc, according to Reform, consisted of three elements: "special interests," the federal government, and Quebec. For Reformers, all three of these elements were inextricably linked. The traditional federal parties — both the Tories and Liberals — were wedded, in general, to an expansion of government and of the welfare state. But these same parties were also guilty of catering to the "illegitimate" demands of Quebecers and "special interests" (unions, human rights advocates, environmentalists, feminists, etc.) in order to curry votes (Harrison et al. 1996; also Laxer and Harrison 1995). Reform successfully (in the eyes of its supporters, at least) blamed this power bloc for Canada's declining economic performance, rising levels of taxation, and escalating public debt, while abridging Canadians' individual political liberties and instituting statist policies in pursuit of social engineering.

Given a significant overlap between Reform and Klein supporters (Stewart 1995), it comes as no surprise that the two parties share similar depictions of the power bloc threatening the people. Yet, there are also some important differences. Reform, after all, never achieved power. By contrast, Ralph Klein's ascension to the premier's office was prefaced by twenty-one years of Tory rule. By the fall of 1992, however, Klein's party was in trouble, beset by a series of scandals involving government loans to business, a declining economy, and rising debts (Taft 1997). In the midst of polls showing that the party would likely lose the next election to the Opposition Liberals, Don Getty stepped down as premier. Ralph Klein succeeded him a few months later (Laxer and Harrison 1995). In the spring of 1993, the new Klein government pulled off a remarkable comeback, winning election handily, a feat repeated again in 1997. How did the Tories do it?

Like the Socreds and Reform, the Klein Tories identified and defined a threat facing the people and named the factors responsible. The threat, said the Klein Tories, was a debt and deficit crisis, fueled by "out of control" government spending (see Taft 1997). The instigators of this crisis, they argued, were: first, special interests similar to those identified by Reform; and second, the federal government, which (in the midst of its own deficit fighting) was off-loading expenditures onto the province. Singled out particularly by the Klein Tories as special interests were Alberta's public unions, especially teachers, nurses, and civil servants, who were significantly "downsized" over the next few years. But university professors, feminists, and gay-rights advocates were not immune to attack either. Klein's own favourite terms of opprobrium for those opposing him have included "communists," whiners," and (more recently) "left-wing nuts." All of those so labelled are contrasted, of course, with the characterization of "ordinary Albertans" as the "real" people (previously discussed). At the same time, the Alberta government renewed traditional calls that greater powers be transferred to the provinces to counteract interference by the federal government in Ottawa. Such demands share common ground with traditional Quebec demand's for a looser federation; hence, the absence of Quebec within the Klein government's construction of the opposing power bloc.

Organization and Mechanisms for Political Participation

Previous descriptions of Social Credit have noted its mix of both authoritarian and plebiscitarian features (Laycock 1990). In contrast to the previous UFA government, in which locals still maintained considerable control over campaign nominations and financing (Monto 1989), power within Social Credit was highly centralized. On paper, local study groups were still the basic unit of organization. In practice, however, Aberhart and the party headquarters determined the party's platform, procedures for nominations, and agenda for conventions and meetings. The candidates who ran for Social Credit in 1935 were "hand-picked" by Aberhart and a few trusted colleagues (Finkel 1989). In Office, Aberhart openly espoused and justified as necessary the practice of patronage (Berton 1991). Policy formulation and implementation was the precinct of technical experts. Social Credit's lone venture into direct democracy — long a staple of Alberta's political culture — was the adoption of the right of recall. The Recall Act was voided, however, when in 1937 Aberhart found himself the first target of the legislation (Finkel 1989; Bell 1993).

Restricted in its actions by the federal government and the courts, beset by growing opposition within, Social Credit turned away from its radical roots. Following Aberhart's death in 1943 and the discovery of oil at Leduc in 1947, Social Credit was gradually transformed under Ernest Manning into a very conventional and conservative government. Party memberships dwindled, the Social Credit League was ultimately dismissed, caucus solidarity was demanded, and subtle — or not so subtle — patronage thrived (Finkel 1989). The becalming of Social Credit's populist impulses was complete.

In many ways, Reform's organizational form reflected Preston Manning's view of the problems that face populist parties and Social Credit's specific experience in dealing with these problems. For Manning (1992), populism has a "dark side" (p. 7) with "racist and other extreme overtones" (p. 25), but can also be "an agent for [positive] political change" (p. 7). Because of populism's dual nature, populist parties, more than conventional political parties, must deal with the construction and operation of their decision-making structures (Manning 1992: 262).

Highlighting populism's positive features, Reform's policies gave priority to the enactment of mechanisms of direct democracy. These mechanisms included "direct consultation, constitutional conventions, constituent assemblies, national referenda, and citizens' initiatives" (Manning 1992: 26), as well as recall (Flanagan 1995: 25). At the same time, Manning is clear that referenda should be used rarely, and only on contentious issues such as capital punishment. Similarly, he believes —

possibly reflecting Aberhart's experience — that the threshold for recall should be set very high.

More fundamentally, Manning does not advocate the abandonment of the role of leadership or of the party executive in favour of undisciplined, bottom-up control. In fact, Manning (1992: 260) has argued that a populist party's greatest vulnerability lies in the fact that its "openness and bottom-up decision-making processes" leave it open to "divisions and attacks from within."

Like Social Credit, Manning and Reform early on addressed the problem of "excess" grass-roots control by highly centralizing its organizational functions. Local constituencies could make policy recommendations, and conventions were relatively uncontrolled compared with many traditional parties. At the same time, Preston Manning and the rest of his brain trust kept tight control over policies and proceedings. As shown in a number of high-profile instances, party members, including elected MPs or constituency presidents, who questioned Manning's leadership or party policy or who damaged Reform's public image, were often publicly ostracized (Harrison 1995a; Flanagan 1995).

Ultimately, these efforts at curtailing or at least shaping Reform's populism failed. The disciples' lack of self-discipline, combined with the leader's wariness of populism's shortcomings, gave way increasingly to authoritarian practices. Reform became the perfect heir to the Socred throne.

By contrast with Social Credit or Reform, plebiscitarianism plays no role in the Klein government. To be sure, the administration makes much of "consulting" with Albertans through its roundtables; the Klein government also polls Albertans regularly, and government policies can be said for the most part to mirror a majoritarian consensus. But the government has consistently rejected the use of province-wide referendums or initiatives, or the implementation of a recall law. While cultivating a populist image, the Klein government's form of organization is in fact best described as "corporatist" (Harrison 1995b). Corporatism is a system of organizing functional interests and influencing public policy that involves the incorporation into society of "members" (individuals, families, firms, or various groups) through a limited number of monopolistic, differentiated, hierarchical, and involuntary associations (Schmitter 1993).

Since its election in 1993, the Klein government has systematically tried to dismantle, or curtail the powers of, groups and organizations opposed to it. For example, the Klein Tories have fought running battles with Alberta Teachers Association, the United Nurses of Alberta, and the Alberta Union of Public Employees, seemingly designed to strip these bodies of their independence.

In their stead, the government has tried to set up a number of intermediate associations subservient to itself, often filled with government appointees. Two examples will suffice: one of the Klein government's first acts after coming to power in 1993 was the concentration of health care under 17 Regional Health Authorities (RHAs) where members are appointed (Taft and Steward 2000); similarly, the government also reduced from 141 to 63 the number of school boards (Kachur and Harrison 1999: xxiii). Not surprisingly, many prominent Tories are RHA members, including former Alberta Treasurer Jim Dinning who oversaw the budget cuts during the first Klein administration and who is now head of Calgary's RHA. The government's power over RHAs was evidenced in 1999 when the entire Lakeland RHA was dismissed by the health minister for running a deficit. But the Klein government also exercises maximum control over elected officials as well, as witnessed by its dismissal — also in mid-1999 — of the entire Calgary public school board which had dared to act too independently. These events are congruent with a corporatist approach to governance.

Conclusion

This article has compared the forms of populism practiced by Social Credit, Reform, and the Klein Conservatives (see Table 2). Clearly, populism remains a mainstay of Alberta politics.

Table 2. A Comparison of Populism in Alberta: Social Credit, Reform, and the Klein Conservatives

	Alberta Social Credit (1935-43)	Reform Party (1987-2000)	Klein Conservatives (1992-2000)
Populist Leader	Rev. Wm. Aberhart (1935-43)	Preston Manning (1987-2000)	Ralph Klein (1992-Current)
Leader's Style	overtly religious personal delegatory	quietly religious technical hands-on	secular personal delegatory
"The People"	Consumers non-elites	Westerners non-elites	stakeholders "ordinary Albertans"
"The Power Bloc"	banks/bankers → Jews big government	special interests federal government Quebec	special interests federal government
Organizational Form/ Mechanisms for Political Participation	centralized/authoritarian plebiscitarian	centralized/authoritarian plebiscitarian	centralized/authoritarian corporatist

Though Ralph Klein and Preston Manning have different political styles, their populist appeal to followers is evident. A degree of continuity can also be seen regarding the three parties' respective characterizations of the people. There is an apparent connection, for example, between Social Credit's "consumers," Reform's concern for middle-class taxpayers, and the Klein government's "stakeholders," all of which stand in marked contrast to broader conceptions of social citizenship. At the same time, Social Credit's and Reform's "non-elite" characterizations of the people are at variance with the Klein government's more generic notion of "ordinary Albertans."

One significant change in Alberta populism is the depiction of the power bloc opposing the people. Social Credit was not a revolutionary movement; nonetheless, some of the party's ideas were radical, a fact that explains in part the stiff opposition from federal government, big business, and mainstream media that initially greeted Social Credit. For Social Credit, "big government" and "big" banks (if not "big capital") were equally problematic. By contrast, Reform's voice was nearly silent on corporate power, while the Klein Tories *are the voice* of both "big government" (at the provincial level) *and* "big corporations."

This change in defining the power bloc is reflected also in organizational practices and mechanisms of populist representation. All three of the parties in question displayed authoritarian tendencies. Indeed, one might argue that Social Credit laid the social and ideological groundwork for anti-democratic tendencies in the province. Nonetheless, early Social Credit also espoused fairly typical populist notions of direct democracy, of which Reform was the obvious heir. By contrast, while the Klein government employs a populist style, built around the personal appeal of its leader, its corporatist practices are at variance with classic populism, actually introducing intermediate bodies between leader and led.

Whither Alberta populism today? In March 2000, as this article was being written, Reform's membership voted overwhelmingly to fold the party into a new political vehicle, the Canadian Alliance Party. In June, the new party elected a new leader, Stockwell Day. The fate of Reform was perhaps foretold years before by Preston Manning (1992: 50–51) when he wrote that populist parties are "more 'human'" than conventional parties; that, in his words, "[t]hey have a lifespan.... They fulfill a purpose"; and that, "when a populist political party is dead, even if it is a governing party, it is really dead and ought to be buried." Neither the demise of Reform, however, nor the eventual defeat of the Klein Tories, will likely diminish Alberta's populist impulses. Populism in Alberta remains as perennial as prairie snows in winter.

Acknowledgements

The author wishes to thank Todd Radenbaugh, Bob Stirling, and the anonymous reviewers of this journal for their insightful comments on an earlier draft of this paper.

Notes

1. The CCF went on to prominence in Saskatchewan, becoming North America's first elected socialist government in 1944. In 1961, the CCF became the New Democratic Party, and has since gone on (at various times) to govern not only Saskatchewan, but also Manitoba, British Columbia, and Ontario.
2. Lone supporters within the media were the Calgary *Albertan* which Finkel (1989: 62) describes as "briefly the 'official organ' of the Social Credit League" and the Ottawa *Citizen* owned by Wilson and Harry Southam who, as Christian Scientists, were Aberhart followers (Berton 1991: 251).
3. This support is not surprising. *Report*'s owner/founder Ted Byfield helped launch Reform in 1987 (Harrison 1995a).
4. Klein has described his supporters as "severely normal" Albertans.
5. Alberta has the lowest tax rates in Canada, including the absence altogether of a provincial sales tax. The Klein government has repeatedly argued that low taxation rates are the primary basis for Alberta's prosperous economy, attracting businesses, capital, and skilled personnel from elsewhere (Laxer and Harrison 1995; Taft 1997). At the same time, Alberta has a large number of user fees and spends less money per capita than many other provinces on public services, including health and education (Taft 1997; Harrison and Kachur 1999; Taft and Steward 2000).
6. The Liberals, 1905–21; the UFA, 1921–35; Social Credit, 1935–1971; and the Conservatives, 1971–today.
7. In 1937, Social Credit introduced the *Accurate News and Information Act*. The Act would have compelled newspapers to publish the government's side on issues, required disclosure of information sources, and threatened suspension of the newspaper or author. Several Alberta newspapers campaigned against the Act, which was struck down the following year by the Supreme Court of Canada (Bell, 1993: 121–22). The Edmonton *Journal* eventually won a Pulitzer Prize for its fight against the bill.

CHAPTER 7

RURAL COMMUNITIES OF THE SASKATCHEWAN PRAIRIE LANDSCAPE

M. Rose Olfert and Jack C. Stabler

ABSTRACT. The rural communities of Saskatchewan's prairie landscape have evolved from a very large number of widely dispersed grain delivery points at the end of the 19th century, through a period of expansion over the first 30 years of 20th century, to a pattern of relatively concentrated population and businesses in an urban-based economy by the end of the 20th century. Technological changes in production agriculture dictated fewer, larger farms and thus a smaller farm population. While the rural economy has diversified beyond its exclusively agricultural base, the new employment generated has not been sufficient to offset the population losses due to agricultural consolidation. Improvements in transportation and communication technology facilitated the concentration of businesses and public sector services in fewer, larger centres. Urban shopping preferences of the rural population have both contributed to, and been influenced by, the concentration of economic activity. While most rural communities have declined continuously over the second half of the century, some rural centres have grown in population, expanded their economic base, and have experienced an increase in their market areas for a limited range of goods and services. In addition to offering a fairly complete range of goods and services, these communities have also become centres of employment for their own and surrounding (farm and non-farm) population. In the 21st century, the viability of small rural communities will depend more on their integration with the centres where economic activity is concentrating, than on their own economic base.

The Settlement Period

Before the arrival of the European settlers with the building of the railway in the 1880s, a few Hudson's Bay Company trading posts appeared in the northern parts of the province. The purpose of these posts was to trade goods of European manufacture for fur collected by indigenous people. The first post in the hinterland was Cumberland House, built in 1774 (Martin and Morton 1938). This was followed by Hudson House (20 miles west of Prince Albert) in 1779, and Manchester House in 1786, about 15 miles north west of Paynton. Limited agriculture developed to contribute to life at the posts. Some of these early trading posts persisted as communities beyond the fur trade era but the fur trade was, in general, not conducive to the development of permanent settlements. Killing the buffalo, allocating reserves for the indigenous people, and the building of the transcontinental railway facilitated the immigration that gave impetus to the establishment of more permanent communities on the prairies.

The small communities covering the prairie landscape today came into existence to serve the needs of the people who originally transformed these plains from grasslands into grain fields and cattle ranches. Their actions completely altered the original ecology, though this was scarcely noted at the end of the 19th century or the beginning of the 20th. The spatial pattern of the trade centre system that developed during the settlement era was influenced both by the early Homestead Acts, which fostered a dispersed pattern of individual farmsteads, and by the production, transport, and distribution technology of the early 20th century.

Under *The Dominion Lands Act* of 1872 a settler could, for a fee of $10, file claim on 160 acres. Title was acquired by satisfying a three-year residency requirement. Although provision was made for settlers to purchase an additional quarter section adjacent to their homestead at a low price, even a half section (320 acres) proved to be too small for establishment of a viable operation in a semi-arid region (Martin and Morton 1938; Mackintosh 1934).

The railroads that were built to carry the region's staple products to world markets provided lines within 10 miles of virtually every farm. When the system had reached its point of maximum expansion in Saskatchewan, farmers could deliver their grain to points spaced at an average distance of only 7.4 miles. Other types of businesses serving the farm trade were encouraged to locate at the grain stops and through this process, the number of communities in Saskatchewan eventually reached 909. Clearly, most of these places were very small (Hodge 1965).

The typical rural community in 1910 would have contained an elevator, railroad station, hotel, lumber yard, livery stable, blacksmith's shop, café, post office, school, bank, butchershop, barbershop, poolroom, churches, a doctor's office, and a community hall. The community's earliest functions were those related to transportation and distribution of the region's export products and the inputs needed for this activity. Gradually some of the inputs were manufactured locally and services to both business and population were provided at the same central locations.

Consolidation and Urbanization

Even as the settlement of the Prairies was nearing completion, however, technological advances were being made that would render obsolete the system that was being built. Only the depression of the 1930s and World War II postponed the inevitable secular adjustment that would have otherwise begun just as the settlement phase ended.

The adoption and implementation of new technologies accelerated during the 1940s and have continued to the present with only occasional and temporary interruptions. By emphasizing mechanization in place of more labour-intensive processes, these new technologies dramatically affected the organization of activities conducted in rural settings and, consequently, the communities that served rural industries and consumers. Particularly important in this respect were the developments that facilitated the substitution of capital for labour in agriculture, thus making it possible for fewer farmers to produce the same, or even a growing, volume of output (Britnell 1939; Barger and Landsberg 1942; Phillips 1956; Fowke 1957). Another factor contributing to the consolidation of agricultural holdings in the Prairie region was the "quota system," under which the amount of grain that could be delivered for marketing was based on the number of acres farmed rather than output per acre. Farmers were therefore encouraged — both by technological progress and by marketing policies set by the Canadian Wheat Board — to expand their land base (Furtan and Lee 1977).

Transportation, communications, and distribution activities were also affected by the development of new technologies. In the 1950s, for example, an ambitious program to update Saskatchewan's inter-city road network was initiated. As a result, paved inter-city roads increased from 750 to 10,000 miles between 1951 and 1971; all-weather, connecting gravel road mileage increased four-fold during the same interval (Saskatchewan Department of Highways records).

One of the first reorganizations to follow the improvement in road access was that of the rural school system. Between 1951 and 1971, some 2,750 schools in rural areas and small communities were closed (Saskatchewan Department of Education records). Grade school students were bused into nearby communities, while high school students travelled to regional high schools that were developed in larger, locally central communities.

Reorganization of the postal and rural telephone systems came next. Between the mid-1950s and 1980, 390 rural post offices were closed and 322 local telephone offices were converted from manual to dial exchanges (Canada Post and SaskTel personal communications). These reorganizations and consolidations eliminated many jobs in rural areas and small communities and transferred others to inter-mediate-sized or large urban communities.

The rural trade centre system adapted to these changes. Smaller communities declined while the larger, centrally located rural communities expanded, at least initially, from the consolidation process. Rural dwellers, both as consumers and producers, contributed to the pattern of concentration. As paved roads were extended into all regions of the province, shopping patterns shifted from the closest rural community to regional shopping centres where more stores, greater variety, and sometimes better quality and lower prices were available. Bypassing of intermediate-sized communities became common, and in response, new commercial development increasingly occurred in the larger centres as it withered away in the small communities (Stabler and Olfert 1992).

Understanding Structural Changes in the System of Communities

Theoretical Framework

Central place theory is the theory most widely used to explain the number, size, and spacing of centres in a system of urban places (Berry et al.). According to this theory, the role of the central place is to act as a service and distribution centre for its hinterland, providing its own, and the adjacent, population with goods and services. Why such functions are provided from central places is explained by the concepts of the demand threshold and the range of the good. The threshold is defined in terms of the minimum level of population and income required to profitably support a particular activity, while the range refers to the maximum area that the activity in question can serve from a particular place. The range is limited because transport costs raise the price of the item as distance from the central place increases. This is true regardless of whether the item is a good distributed from the centre or is one that customers have to travel to the centre to obtain.

Since the threshold and range will differ among various activities, a hierarchical spatial structure results in which the activity with the lowest-threshold requirement is found in all central places. In today's context, a gasoline service station would typify a service function with a low-demand threshold. Only a small population is required to provide the level of demand necessary to support a gasoline station: many therefore exist, and they are distributed widely — wherever a small concentration of population is found. Activities requiring a larger threshold, such as furniture stores, are found in fewer and larger places. Since the size of service areas varies directly with the size of centres, the complementary regions of small places are contained within those of larger places.

The required number of functions of each type, and thus the number of centres of each size within the system, is largely a function of total population and income, while the spacing of centres is determined by population density and accessibility. Higher incomes and larger populations are associated with a greater number of functions. Although the number of centres is directly related to population density, it is inversely related to the quality of the region's transportation systems. Each successively higher-order function, offered from increasingly higher-level centres, is provided to all lower-level centres and the rural populations within the ever-larger market areas of the higher-order functions. Often, several functions will have approximately the same demand threshold and a similar range (Wensley and Stabler 1998). The highest level of centre will offer all functions including those of the highest order. The lowest level centre will offer only the lowest order of function.

Overview of Changes in the Trade Centre System, 1961–1995

Saskatchewan's communities were classified into functional levels, in the Central Place Theory context, for 1961, 1981, 1990, and 1995. These comparisons are provided in Table 1. Between 1961 and 1981, a very substantial downward movement of communities in the middle categories occurred. In 1961, for example, there were 317 communities situated in the three tiers between the Secondary Wholesale-Retail (SWR) level and the lowest functional classification, the Minimum Convenience Centre (MCC). By 1981, the number occupying this interval had decreased to 188. During this time period, the number of communities in the Complete Shopping Centre (CSC) category decreased from 29 to 22 (25 percent), but the Partial Shopping Centre (PSC) and the Full Convenience Centre (FCC) categories experienced much greater downward movement.

Between 1981 and 1990 there was some further downward movement of centres in the middle categories, but at a much slower rate. The number of communities in the three clusters between the SWR and the MCC levels declined from 188 in 1981, to 169 in 1990. What is most striking about the latter period is the pronounced decline of communities that had CSC status in 1981 — only 6 of 22 centres (27 percent) remained in this category in 1990.

In the most recent period, 1990–1995, there was again a very substantial downward reclassification of communities in the middle categories. The number of centres between the SWR and the MCC levels decreased to 88. Thus, over the 34 years included in this study, middle-level communities decreased from 53 percent to 15 percent of the total, while the number of centres in the bottom (residual) category, MCC, increased from 45 to 84 percent.

The most recent five years were again characterized by stability at the top three functional levels; indeed, there was one addition to the CSC category. The most striking feature of the 1990–95 period, however, was the dramatic downward reclassification of the communities that had occupied the PSC and FCC categories in 1990: 50 percent of the 1990 PSCs and 70 percent of FCCs had moved to a lower classification.

Table 1. Functional Classification, Saskatchewan Centres, 1961–1995

Functional Classification	1961	1981	1990	1995
Primary Wholesale-Retail (PWR)	2	2	2	2
Secondary Wholesale-Retail (SWR)	8	8	8	8
Complete Shopping Centre (CSC)	29	22	6	7
Partial Shopping Centre (PSC)	99	30	46	22
Full Convenience Centre (FCC)	189	136	117	59
Minimum Convenience Centre (MCC)	271	400	419	500
Total	598	598	598	598

Source: Stabler and Olfert, 1996.

The 1995 profile of the six functional classifications, provided in Table 2, indicates the role played by communities at each level. Communities in the top four functional classifications are distinguished from those in the bottom two in that they provide a full range of producer and consumer services, as well as a nearly complete range of common infrastructure.

The variation in functional roles among the top four is characterized by depth and variety. PWR centres, for example, have multiple outlets of each consumer function and therefore offer the widest range in product and service variety, quality, and price. In the next two levels, SWR and CSC, virtually all consumer functions are present, but there are fewer outlets of each type and thus a more limited range of choice. PSCs, while still offering most consumer functions, provide only a very restricted range of options. Further, some higher-order functions, such as furniture stores and laundries/dry cleaners, are much less common at the PSC level. A final important distinction is that many national and international trade and service franchises and chains have a major presence in communities in the top three categories. Although they are occasionally present in PSCs, they are completely absent below that level (Wensley and Stabler 1998).

The top four categories are also characterized by the presence of a range of producer services and infrastructure, with both numbers and variety increasing at higher functional levels.

There is a very substantial difference, however, between the PSC and the FCC classifications. FCCs do not provide a full range of consumer services, and while several producer services are present, business services are largely absent at this level. Further, farm equipment outlets at the FCC level are more likely to represent only the sale of parts rather than implements. Infrastructure, too, especially hospitals, becomes less common at this level. Nevertheless, most FCCs continue to perform a locally useful function in the provision of several day-to-day requirements. They provide a service, in a rural setting, akin to what 7-Eleven stores offer in urban space.

Finally, the 500 communities in the MCC category, as a group, no longer perform a coherent role in the trade centre network. There is no single function that can be counted on to be present in these places. While the average MCC will have approximately four consumer outlets, they consist of an eclectic combination of functions that vary from place to place. Infrastructure, except for grain elevators, is almost totally absent and only the occasional producer or producer service outlet is found in communities at this level.

The pattern of urbanization evident in the changes described above is part of a continuing process of urbanization that extends to all of North America. The population of the 598 subject communities was 538,666 in 1961 (approximately 58 percent of the province's total). By 1981 there were 679,622 people living in these centres (approximately 70 percent), and by 1995 the population of these communities was 728,675 or 72 percent of the provincial population. Saskatoon and Regina combined accounted for 22 percent in 1961, increasing to 36 percent in 1995. Business outlets also became more concentrated in the largest centres, with

Table 2. Average Population and Number of Businesses of Various Types in Saskatchewan Trade Centres, 1995

Type of Business	Minimum Convenience (n = 500)	Full Convenience (n = 59)	Partial Shopping (n = 22)	Complete Shopping (n=7)	Secondary Wholesale-Retail (n=8)	Primary Wholesale-Retail (n=2)
Population	179.61	860.68	2,049.73	4,808.86	17,609.88	184,235.50
All Consumer	3.72	18.73	43.09	90.14	227.00	1,640.00
General Store	0.29	0.90	1.05	1.86	3.13	12.00
Grocery Store	0.39	1.59	3.23	3.71	11.00	49.50
Special Food	0.09	0.80	1.41	3.29	5.38	42.50
Auto Sales	0.14	1.15	3.41	5.71	15.88	63.50
Gas Station	0.37	1.34	2.64	4.86	13.00	62.50
Clothing Store	0.06	1.08	3.32	7.57	12.38	82.00
Furniture Store	0.02	0.25	0.64	2.43	4.25	25.00
Home Furnishing	0.06	0.53	1.64	5.29	13.50	82.00
Restaurant	0.32	1.64	4.27	8.86	23.88	
Drug Store	0.04	0.81	1.59	2.43	4.88	31.50
Special Retail	0.12	0.98	3.82	9.57	21.50	182.00
Credit Agency	0.47	2.02	4.05	11.14	39.63	
Hotel	0.39	1.05	2.64	4.29	7.50	22.00
Laundries/Dry	-	0.10	0.23	0.43	2.25	18.00
Personal Services	0.02	0.12	1.05	2.00	6.38	36.50
Auto Repair	0.25	1.24	2.23	4.71	14.63	105.00
Car Wash	0.16	0.98	1.86	5.57	16.63	113.50
Recreation	0.03	0.41	1.18	2.00	3.75	39.00
Bank or Credit Union	0.50	1.73	2.86	4.43		62.00
All Producer Services	0.98	5.31	11.09	21.57	64.75	750.00
Warehousing	-	0.08	0.09	0.43		
Farm Equipment	0.22	1.10	2.55	2.86	6.25	24.50
Bulk Fuel	0.30	1.29	1.86	2.43	4.25	10.00
Wholesale	0.25	1.14	3.32	8.00	31.63	435.50
Building Materials	0.18	1.39	2.23	4.57	8.38	39.00
Business Services	0.04	0.31	1.05	3.29	13.63	232.50
All Producers	1.22	6.29	14.32	32.71	101.38	970.50
Construction	0.65	3.53	7.82	16.14	52.38	522.00
Manufacturing	0.29	1.47	3.95	9.29	23.75	283.00
Transportation	0.28	1.29	2.55	7.29	25.25	165.50
Doctor*	0.07	0.73	0.95	1.00	1.00	1.00
Hospital*	0.02	0.41	0.91	1.00	1.00	1.00
Special Health Care*	0.02	0.88	1.00	0.86	1.00	1.00
High School*	0.29	0.98	0.95	1.00	1.00	1.00
Grain Elevator*	0.64	0.93	0.91	1.00	1.00	1.00

Source: Stabler and Olfert, 1996.

*For these variables the percentage of communities offering selected services/facilities is shown.

Saskatoon and Regina increasing their share of all business outlets from 19 percent in 1961 to 40 percent in 1995.

This consolidation of population and businesses and the downward filtering of the communities are explained historically by: the decrease in the number of farms (from 103,000 in 1956 to less than 57,000 in 1996) and the subsequent depopulation of rural areas; the geographic extension of rural dwellers' shopping patterns due to a greatly improved inter-city highway network; the consolidation of rural schools made possible by the same improvements in the road network; and the growth of urban-based service sector employment. In the 1980s and 1990s the consolidation of public infrastructure added to the shift away from smaller, more dispersed communities into fewer, larger, well-situated rural centres. The location patterns of national and international chain stores and franchises have also added to the viability of larger communities, enhancing their attractiveness relative to the smaller places that these businesses avoid (Stabler and Olfert 1990).

The map in Figure 1 shows the spatial arrangement of the 39 communities that occupy the top four functional classifications in 1995 (29 black triangles for the Shopping Centres and 10 black squares for the Wholesale-Retail centres). The experience of shopping centres in 1961 that have subsequently lost that status is also shown. While the 1995 numbers in the top four categories are fewer than at any time in the past, the spatial distribution is still such that virtually the entire southern half of the province is within 30 minutes driving time of one of these centres (except for the southwest corner).

While the general pattern of trade centre change was one of consolidation over the observation period, there were also differences within the province. The viability of communities will be determined in part by the level and stability of the income of the population in the market areas surrounding each centre. These income levels and population densities vary considerably by agricultural area. In a study of differences in income levels, agricultural sales, off-farm employment, and the diversification of the agricultural sector by Crop District, Stabler and Olfert (1994) defined three Types of agricultural areas, Type 1 with predominantly positive growth indicators (relative to the provincial average), Type 2 with predominantly negative indicators, and Type 3 with mixed indicators.

Over the 1961–95 period, communities in the Type 1 area, the north, fared better than those in Type 2 (the south-southwest), and the former retained a larger share of the province's most viable communities — those in the top 3 levels of the trade centre hierarchy. These results confirm that a more diversified economic base provides a more hospitable environment for community development.

Several factors have influenced the restructuring of Saskatchewan's trade centre system during the past five decades. Technological improvements in agriculture, transportation, communications, and distribution activities, coupled with changing tastes, have all contributed, in varying intensities at different times, to the consolidation that has taken place from the end of the settlement era to the present.

During the 1980s and 1990s, urbanization led to a growing share of the province's population living in its cities, towns, and villages. Increasingly, the growing urban population chose to live in Saskatoon, Regina, and a couple dozen larger communities at the top of the trade centre hierarchy. Businesses followed, or accompanied, the shift in population, and this pattern was reinforced as the shopping patterns of rural dwellers took on an increasingly urban focus.

Continuous improvements in technology in the agriculture sector have contributed to the long-term decline in the real (inflation-adjusted) price of grains and many other agricultural products. Input costs, on the other hand, have not declined in keeping with agricultural prices. The termination of the Crow transportation subsidy in 1995 represented a substantial increase in transportation costs to western farmers producing for export markets. As a consequence of these unfavourable changes, net farm income, which accounted for two-thirds to three-quarters of farm household income through the 1960s and into the mid-1970s, had fallen to only 28 percent of farm household income in 1995. Income from off-farm

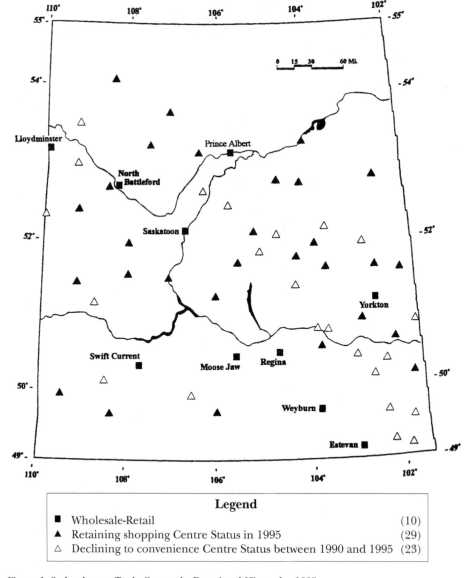

Legend
■	Wholesale-Retail	(10)
▲	Retaining shopping Centre Status in 1995	(29)
△	Declining to convenience Centre Status between 1990 and 1995	(23)

Figure 1. Saskatchewan Trade Centres by Functional Hierarchy, 1995.

jobs had become the single most important source of support for farm families, accounting for 48 percent of farm household income. Transfers and investment income accounted for the remaining 24 percent.

Adaptations

Clearly, the historical pattern has been one of many communities losing functions and population, and a few gaining market area and becoming stronger as a result. This process strongly favours the largest centres in the province, along with a few well-located regional centres.

Much has been made of the new opportunities that the information age and associated technology afford rural areas. Technically, advances in communications, data processing and transmission technology make rural locations possible for the rapidly growing services sector reliant on this technology, as well as other businesses where the new technology can reduce the barriers imposed by remoteness. In the case of Saskatchewan, however, the services sector has remained highly concentrated in the province's largest centres. In addition to technical feasibility, there has to be a willingness on the part of the entrepreneur and/or employees to pursue the activity from a remote setting. To this time a few remote, but idyllic, locations have been the primary beneficiaries of this technology. Thus, there is little reason to believe that the new technologies will reverse the fortunes of rural Saskatchewan.

Multi-Community Co-operation and Regionalization

Not every community in Saskatchewan can offer anything like a full range of goods and services. One means of accessing a broad range of private and public sector functions is through cooperation among communities (Stabler 1996). The area over which such cooperation is feasible may be thought of as a functional economic area, i.e., an area large enough to include common places of residence, work, and shopping of the inhabitants. Regional structures based loosely on this notion have proliferated in Saskatchewan in the 1980s and 1990s. Some of the more important examples include: Intermunicipal agreements, Regional Economic Development Authorities, and Health Districts.

Intermunicipal agreements among Saskatchewan's 497 urban municipalities (excluding resort villages) and 297 rural municipalities are common for the selective delivery of services, sharing of equipment and facilities, and jointly managing programs. A minimum population size is frequently implied for the construction of a state-of-the-art land fill site: for example, the purchase of new equipment, or bulk ordering of supplies. With declining populations in many rural areas and in smaller centres, an individual municipality, urban or rural, may need to share the costs of new purchases or projects with other municipalities. Some provincial grants are tied to inter-municipal cooperation and, in general, provincial policies and legislation have been supportive of these initiatives.

A survey in 1992 (Saskatchewan Community Services 1992) showed that the average urban municipality in Saskatchewan provided 6 services on an intermunicipal basis and that most often 2 to 4 municipalities were members of such an agreement. The arrangements are usually in the form of written agreements, and funding is mainly by municipal contributions, grants and user charges. The most common inter-municipal services noted in the survey were fire protection, regional libraries, union hospital districts, ambulance and emergency planning.

Other services for which intermunicipal agreements were reported included recreation, landfill sites, care homes, regional parks, animal/pest control, rural development corporations, road maintenance, senior low income housing and municipal administration. The potential for co-operation in a wider range of activities and including a larger number of municipalities is substantial, especially as population size decreases in many areas and as the services themselves are more demanding in terms of the minimum population required to justify equipment and/or technology.

Regional Economic Development Authorities (REDAs) were introduced in Saskatchewan's Partnership for Renewal economic strategy in 1992 (Saskatchewan Economic Development 1996). Encouraging and attracting economic development initiatives requires a local labour force, support services, and facilities which may be beyond the means of an individual municipality to provide. A combination of villages, towns, rural areas, and cities, however, may represent an attractive package for new business development as well as having the required resources, human and otherwise, to support infrastructure and business development.

REDAs allow communities and organizations to jointly make plans for economic development in their region. The provincial government provided some cost-shared funds to help with the formation of REDAs and to help established REDAs build their service capacity and form partnerships. In 1999 there were 27 REDAs in the province, some of which were reporting successes in their economic development efforts.

Health Districts were created following the Saskatchewan Visions for Health Reform released in 1992. Under the Health Districts Act, the guidelines for the formation of a District were that it consist of a contiguous land area with a total population of no less than 12,000. The Districts were to be formed on the basis of the location of communities, population distribution, geographic barriers, trading and commuting patterns, location of current health facilities, and population health status (Kouri 1999). This process of regionalization of health services delivery represented a centralization of powers from more local and facility-based boards, as well as a decentralization of power from the provincial government.

Thirty such Health Districts were formed through voluntary associations of component municipalities. Within these boundaries, the various health services groups were amalgamated. Funding comes to the Districts from the provincial government determined by a needs-based formula reflecting the size and composition of the population and the locations at which the population receives services. In 1996, the 30 Health Districts in southern Saskatchewan ranged in size from a population of 10,215 (Gabriel Springs Health District) to 225,370 (Saskatoon). Four of the Districts have populations below the initially prescribed 12,000.

Regionalization and multi-community organizations and agreements indicate recognition of the obsolescence of the individual community as the focal point for economic development and service delivery. The need for a regional focus reflects both the declining populations in many rural areas and small communities, and the changing requirements of service delivery and economic development. Larger, more specialized machinery and equipment, the demand for state-of-the art services and facilities, and improvements in transportation and communication have all led to increasing population requirements for many services or activities. At the same time as the need for a regional approach is widely recognized, cooperation

among communities is compromised by the fact that the benefits of that cooperation are shared unevenly among communities and not necessarily in proportion to
the costs incurred.

Commuting

In addition to attracting a growing share of the population, the province's
urban centres also provide an increasing amount of employment for people who
reside elsewhere but commute into these centres. Commuting is important in rural
areas for both the farm and non-farm populations. For farm families, off-farm
employment necessarily entails commuting, sometimes to another rural location
and sometimes to an urban centre. The non-farm population in small centres or
rural areas may also need to commute to find employment, especially where there
are two income-earners in a family.

To examine commuting patterns, the largest 62 centres were taken as the
potential sites of urban employment and therefore potential commuting destinations (Stabler, Olfert and Greuel 1996). Then, for 1981, 1991, and 1996 Censuses,
special tabulations on place-of-residence and place-of-employment were used to
define employment patterns. Anyone who works some place other than where they
live is considered a commuter. Both commutes to one of the 62 centres from
another location and commutes from one of the 62 places to a rural area or to
another urban area are counted as commutes.[1]

Given this representation of commuting, there was an increase in the amount
of commuting between 1981 and 1991 of 6,250 (from 38,585 to 44,835). In percentage terms this represents a 16 percent increase over a period when the total
employment in the province increased by only 6 percent. Again between 1991 and
1996 the number of commuters grew by 16,940, an increase of 38 percent. Total
employment grew by less than 1 percent between 1991 and 1996, clearly indicating
a growing dependence on commuting.

The distribution of commuters between *to urban* centres and *from urban* centres
changed as well. Overall, there was a very large increase in *to urban* commuters and
a decrease in *from urban* commuters between 1981 and 1991, showing that the
increase in commuters was more and more made up of rural residents commuting
to urban centres of employment rather than urban residents commuting out to
employment in rural areas. From 1991 to 1996 there was an increase in both *to
urban* and *from urban* commuters, with commuters *to urban* centres growing by just
over 10,000, and commuters *from urban* centres growing by just under 7,000.

While the pattern of increased commuting was evident for most parts of the
province there were, in all years, some rural municipalities (RMs) from which there
was virtually no commuting to selected centres of employment. These RMs were not
associated with any urban-based labour market area (LMA). Clearly, for some parts
of the province commuting to urban centres for employment is not an option.

Finally, the distribution of commuters by gender reveals that nearly all the
increase in commuting between 1981 and 1991 was accounted for by females (96
percent). In both years there were more male than female commuters, but the
gap narrowed considerably by 1991. Between 1991 and 1996, the increase in
female commuters again outnumbered the increase in males, but females
accounted for a smaller percentage of the increase (52 percent) than in the previous interval. Females made up 38 percent of the commuters in 1981 but

accounted for 48 percent in 1996. Since females have a lower participation rate (i.e., are less likely to be in the labour market) than males, this also means that a greater proportion of the female labour force (compared with males) is commuting.

Among the 62 communities there is considerable variation in the amount and relative importance of in-commuting. For the largest centres, Saskatoon and Regina, commuters supply less than 10 percent of the urban centre's labour force, while in some of the smaller centres (population of 1,000 and less) this proportion increases to close to 50 percent. The absolute number of commuters, however, is greatest to the largest urban employment centres.

Average distances travelled to places of employment also vary considerably with the size of the employment centre. On average, the largest centres draw from an area with a 40-mile radius, while the smallest employment centres draw from an area with a 23-mile radius. Commuting distances have increased by more than 10 percent since 1981.

Off-Farm Employment of Farm Households

The widespread commuting referred to above includes commuting by farm families to places of employment in nearby communities. In many cases farm families are dependent upon larger rural communities for their employment opportunities, as well as for a place to obtain goods and services.

Farm families have turned increasingly to off-farm jobs and non-farm businesses for support. In 1996 more than two-thirds of the female farm labour force, and more than one-third of farm males, spent over 50 percent of their working hours at jobs other than on their farm (Stabler and Olfert 1999). This integration of the farm labour force into the non-farm economy involves a substantial amount of commuting. Thirty-four percent of the farm labour force commutes to work, primarily in the largest communities. Female commuters outnumber males in absolute terms and even more so in terms of the percentage of the (gender-specific) labour force that commutes.

Age-Selective Net Out-Migration

Population movements from rural to urban centres and from smaller rural centres to larger communities do not affect all age groups equally. Centres of growing employment will attract population in the labour force ages. A high proportion of the population in the 25-44 age groups also implies a large proportion in the <15 age groups. Rural areas or small communities with limited and declining employment opportunities will have a high proportion of the population in the 65+ age groups. Centres too small to provide even the minimum conveniences to the elderly population will, however, not be able to attract the 65+ population.

Table 3 illustrates the differences in the age structure of the population in a Primary Wholesale-Retail centre (Saskatoon), a Partial Shopping centre (Kamsack), and a Full Convenience Centre (Eatonia). While the age structure will vary among communities at each level of the hierarchy, the same general pattern applies. The largest centres are the primary sites of economic activity and, as such, attract population in the prime labour force ages (25-44). Larger rural communities heavily dependent on traditional agriculture show a relative absence of the 25-44 age groups but are attractive destinations for the 65+ age groups. The smallest centres, such as Eatonia, are less attractive as retirement destinations as they have fewer services.

Table 3. Age Distribution of the Saskatchewan Population by Hierarchical Level, 1998

Age Group	PWR (Saskatoon)		PSC (Kamsack)		FCC (Eatonia)	
	#	%	#	%	#	%
<15	44,268	21.49	464	18.72	128	21.19
15-24	29,519	14.33	258	10.41	78	12.91
25-44	65,979	32.03	566	22.84	145	24.01
45-54	24,060	11.68	235	9.48	51	8.44
55-64	15,923	7.73	196	7.91	57	9.44
65+	26,264	12.75	759	30.63	145	24.01
Total	205,992	100.00	2,478	100.00	604	100.00
Elderly Dependency ratio	0.1939		0.6048		0.4381	

Source: Saskatchewan Health. 1998. *Covered Population.*

The age-selectivity of out-migration from rural areas and smaller rural communities is summarized by the high elderly dependency ratios, especially in the larger rural centres. In the examples shown in Table 3, Saskatoon has an elderly dependency ratio of .1939, showing that for every person in the potential labour force (ages 15-64) there are .1939 persons over the age of 65. Viewed in another way, for every person over the age of 65, there are 5.16 people of labour force age. For Kamsack, there are 1.65 persons of labour force age for every person over the age of 65, and in Eatonia, 2.28. Typically the highest elderly dependency ratios (the lowest proportion of 15-64 relative to 65+) are found in the Partial Shopping Centres (like Kamsack), indicating they are still attractive retirement destinations, although there is not sufficient economic activity to attract or retain those of labour force age. The need for schools, recreation facilities, health services, business outlets of all kinds, and public infrastructure is affected by the age structure of the resident population.

The Aboriginal Population

Throughout the first 60–70 years of the twentieth century, the aboriginal population played virtually no role in the development of the trade centre system described above. The policy of setting aside reserves for Registered Indians largely excluded them from the communities developed to serve the agricultural population. The reserves themselves have always served as communities for the Registered Indian population, of course, though little is known about their limited trade centre activities or how these evolved over time. The very nature of the reserve meant that population growth was dictated by natural increase rather than by net migration patterns as has been the case for the rest of the province. Facilities on reserve were more often the result of decisions by the federal Department of Indian and Northern Affairs than by the private business sector. Even when the Registered Indian population moved to urban centres, however, they were not visible in the published data sources. The population data on which much of the analysis reported here is based, for example, do not identify the geographic location of the Registered Indian population, only their band affiliation. This did not change until 1998.

People of aboriginal ancestry, though not Registered Indians, were always present in some southern communities, but they were relatively few in number and not formally distinguishable from the rest of the population in conventional data sets.

During the second half of the 20th century, several factors changed the profile of the aboriginal population in the province's communities. With a high rate of natural increase and limited economic opportunities on reserve, native people began moving into urban centres. For some centres such as Prince Albert and North Battleford, the presence of a rapidly growing aboriginal population has significantly increased their growth rates. For some others, population declines in the non-native population have been offset to some extent by the growth of the aboriginal population.

The reserves continue to be the communities of reference for much of the aboriginal population, and an expanded resource base through settlement of Treaty Land Entitlements may increase their ability to serve in this capacity. However, since few of the trade centre functions, or public services are available on reserves at present, the aboriginal population forms an important part of the 'market' for the private and public services in nearby non-reserve communities. In addition to the reserve communities, however, it is clear that many other Saskatchewan centres, especially the larger urban places, will become home to the aboriginal population. A high rate of natural increase in this population, combined with a declining non-native population in all but the largest cities, will continue to increase both the absolute and relative importance of the aboriginal population in the future of many Saskatchewan communities.

Rural Communities in the 21st Century

In the 21st century rural communities in Saskatchewan will continue to be affected by many of the same forces prevalent in the 1990s. Further farm consolidation appears inevitable, resulting in fewer farms and fewer farmers. Communities will also likely decline in number, with a small number growing and increasing their regional importance and function, while most communities lose population. It is likely that economic activity and population will continue to concentrate in the major urban centres, with a limited number of larger rural communities performing a narrower range of functions. A few larger rural communities may thrive in their role of regional shopping centres and points of distribution for a limited subset of public and private services and functions. Some of the smallest communities will cease to exist; others will remain as places of residence for a time, while continuing to lose their business and public services functions.

This pattern of further consolidation will increase the integration (and dependence) of the rural population and the smaller rural communities with those centres that are attracting economic activity and population. In addition to providing employment for their resident populations, the rural growth centres also become focal points for employment for the surrounding farm population, other rural population, and populations of smaller centres. The resulting commuting and trading area patterns form *de facto* functional economic regions. Municipal government, however, remains structured much as it was in the early 1990s.

Multi-community cooperation within these regions, and some measure of coordination and planning, may be a means for participation by, and survival of, individual communities within the region. Accessing regionally provided services and facilities will be important to all participants, regardless of the specific location of services within the region. As many individual communities lose functions, a next-best alternative will be to be able to access these services regionally. Larger functional economic areas will facilitate this process.

Notes

1. Since the focus was on the 62 centres, some commutes will not be detected. If a rural dweller works in a nearby small town that is not one of the 62 centres or in a nearby potash mine or oil and gas field, he/she will not be counted as commuting to urban centres.

CHAPTER 8

CHANGING RAILROAD LANDSCAPES
ON THE PLAINS AND PRAIRIES

Alec Paul

ABSTRACT. The northern Great Pains, including the Canadian Prairies, were opened for commercial agriculture by the spread of railways. In the process a characteristic railway landscape or "railscape" was developed in the region. Despite the inroads of the truck and the automobile, much of this railscape still survived in 1970; it was based on an extensive network of branchlines serving large numbers of small wooden grain elevators. By the year 2000, however, major changes in the railscape have occurred and more were on the horizon. The rail net shrank considerably, and the elevator rows along the tracks that had anchored many small towns and villages largely disappeared. These landscape changes are reviewed here over the period 1970 to 2000, organized in terms of main-line landscapes, other active railscapes, and railscapes of abandonment.

Introduction

Railways transformed the Great Plains of North America, allowing accelerated exploitation by the capitalist economic system even as they crossed the region. Impacts on the landscape were significant in the advance of the rails in the sixty years from 1869 — the completion of the first transcontinental — to the economic crash of 1929, and they still are in the year 2000. The role of the railway has shifted from promoting settlement and agriculture to today's long-distance movement of large volumes of a few basic traffics such as coal, grain, containers and trailers (intermodal). Thus over much of the northern Great Plains, the railway is in retreat. Further, as government regulation has diminished, so has the railnet. Long stretches of track already have disappeared, leaving a select few ultramodern high-speed, high-volume main lines of the Class 1 companies. A Class 1 railroad is a major company, defined by the U.S. Interstate Commerce Commission on the basis of annual operating revenue exceeding $256 million (Lewis 1996). A list of the company acronyms used in the current article is given in Table 1. The Burlington Northern Santa Fe (BNSF), Union Pacific (UP), Canadian Pacific (CP) and Canadian National (CN) have emerged to dominate rail transport in the northern Plains. At the same time, a number of so-called "short" and "regional" lines help to maintain rail service to the largely agricultural outback, an area that is no longer the target of the main line companies. However, some of the regional lines are healthy enough to fight for a small share of the lucrative high-volume traffic. For

Table 1. Railway Company Acronyms

BN	Burlington Northern
BNSF	Burlington Northern Santa Fe
CN	Canadian National
CNW	Chicago & North Western
CP	Canadian Pacific
CRIP	Chicago Rock Island & Pacific
DM & E	Dakota Minnesota & Eastern
DMV & W	Dakota Missouri Valley & Western
GN	Great Northern
MIL	Chicago Milwaukee St. Paul & Pacific
NP	Northern Pacific
RRV & W	Red River Valley & Western
SOO	Soo Line
SRC	Southern Rail Co-operative
UP	Union Pacific

example, Dakota, Minnesota and Eastern (DM & E) is trying to extend its tracks to the Powder River coalfields of Wyoming, at present the exclusive preserve of BNSF and UP.

This article concentrates on changes in the railway landscapes of the northern Great Plains from Nebraska, the Dakotas, Wyoming and Montana south of the 49th parallel to the Canadian Prairies of Saskatchewan and Alberta to the north. Over the past thirty years, new railway landscapes have emerged in this region while others are being re-made and redefined. The reasons behind both rail construction and abandonment are complex but intimately linked with current social landscapes. Here, a case study approach is adopted that verbally and photographically assesses the current railscapes in this region while making suggestions about ongoing shifts for the next decade.

Railscapes of the Present

In the early fall of 1999, the railroad at Newcastle, Wyoming, where the High Plains meet the Black Hills, seemed a world away from the scene photographed at Minton, Saskatchewan a few days earlier (Figures 1 & 2). Newcastle has a double-track mainline with heavy steel rail on concrete ties. Along these tracks rolls a seemingly endless series of BNSF coal trains each stretching over a mile. They originate in Wyoming's extensive Powder River coalfields then travel east to help to meet the United States' seemingly insatiable appetite for electrical power. Minton, on the other hand, was the terminus of a grain branchline recently abandoned by the Canadian Pacific, a relic of an earlier era on the plains. Only the wooden ties and two old grain elevators remained. The lightweight rails had been lifted, leaving only a graded hiking trail through a quiet countryside. For the residents of both Newcastle and Minton, this changing railscape is something of a double-edged sword.

Alec Paul

Figure 1. Minton, Saskatchewan: recently abandoned CP grain branch, 22 September 1999.

Alec Paul

Figure 2. Newcastle, Wyoming: northbound BNSF coal empties, 27 September 1999.

Newcastle, with a sawmill and small oil refinery, has far more trains than it needs or wants. The majority of them pass through without stopping, their whistling and interruption of traffic creating a disturbance that forces motorists to detour over the new "viaduct" built to accommodate the increased rail traffic. Meanwhile farmers around Minton find themselves with the opposite problem Without a railway the road network is their only option for transporting grain. As the roadways deteriorate due to the pounding of heavy grain trucks, the Rural Municipality and the provincial government must now search desperately for ways to finance repair and upgrading.

Large areas of the northern Great Plains are now bereft of rail service. A few grain elevators remain on abandoned lines, but they are now only served by truck. High-volume concrete or multiple-steelbin elevators commonly known as "subterminals" or "inland terminals" occupy strategic locations on surviving rail lines. They are far bigger but fewer in number than the old wooden elevators, and may not even be located in or close to a town or village. Thus, the changes in the grain collection system mirror the shrinking rail net. To investigate the antecedents of today we must look into the past.

The Antecedents

Railroads expanded on the northern Great Plains up to the stock market crash of 1929. Despite the inroads of the automobile and the truck there was little railway retrenchment through the Great Depression and World War II period. In the 1950s, rail passenger numbers declined as highways improved and airline passenger traffic mushroomed. In addition, the technological change from steam to diesel on the railways drastically reduced railway employment in the 1950s. Coupled with declining freight revenues in the 1960s, these factors triggered a dramatic upheaval in railscapes in the 1970s. Thus, 1970 was a turning point for the changing railscapes described in this article.

Around 1970 numerous main lines still stretched across the northern Great Plains (Figure 3). CN and CP traversed the Canadian prairie provinces, and the CP's Soo Line ran southeast from Moose Jaw, SK to Chicago, IL. Three railroads crossed the Dakotas and Montana: the Great Northern (GN), Northern Pacific and Milwaukee (for Chicago, Milwaukee, St. Paul and Pacific), all with main lines out of Chicago running to the Pacific Northwest. Further south, the Chicago and North Western (CNW), Burlington (Chicago, Burlington and Quincy) and the UP cut across Nebraska and Wyoming. Over much of the region a dense net of branchlines serving chiefly to ship grain outwards was still operated by the Class 1 carriers. Points where surrounding farmers could deliver their grain to wooden elevators had been established every 10–15 km along these branches earlier in the century. Although some of these places and elevators had already disappeared (Williams and Everitt 1989), in 1970 much of this branchline railscape was still intact.

The thirty years since 1970 have seen substantial modifications to these railway landscapes. A major influence on the northern U.S. plains has been the rapid rise of Wyoming to become the number one coal-producing state in the U.S. Much of this coal is from the Powder River fields around Gillette and is bound for electricity-generating stations of the U.S. mid-section. It is no coincidence that all this coal is moved out of the Powder River country by the two surviving "mega-railroads" of the U.S. West, Burlington Northern Santa Fe (BNSF) and UP. BNSF resulted from an

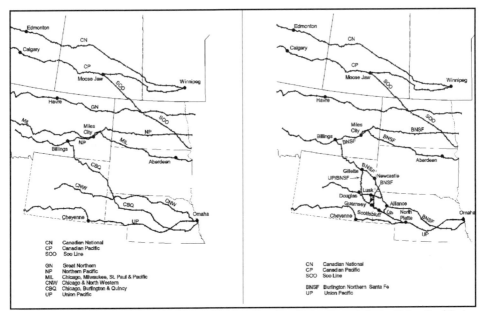

Courtesy Information Systems Division, Canadian Plains Research Center, University of Regina
Figure 3. Main lines on the northern Plains and prairies, 1970 and 2000.

amalgamation in 1970 of the Burlington, GN and NP to form the Burlington Northern (BN), which merged with Santa Fe in 1995. UP took over the Chicago and North Western in 1996.

At the beginning of the new century, the geography of railways on the northern Great Plains is vastly different from that of thirty years ago (Figure 3). Grain shipment also experienced drastic changes during 1970-2000 in Canada and the U.S. The multitude of smaller and older wooden elevators rapidly gave way to subterminals on the northern U.S. plains in the 1970s. At the same time many of the grain branchlines were abandoned. The Staggers Act of 1980 that effectively deregulated the American rail industry (Sorenson 1984), led to the proliferation of shortline companies which took over much of the remaining branchline system from the large railroads. Still, farmers in many areas faced increased trucking costs to market their grain as the number of railway elevator points plummeted.

On the Canadian Prairies, rail deregulation was slower. The federal subsidy to grain movement by rail in Canada (the "Crow Rate") was removed by Ottawa in 1983. However, a program to soften the impact of the demise of the "the Crow" continued until crop year 1994/95. So too did the process of closely regulated abandonment by the Canadian railways. By this time, the cushion provided by the Canadian government was now gone and western Canada faced the same scenario that had confronted the U.S. Plains fifteen years earlier. Thus, since 1995, elevator numbers and railway mileage have dropped dramatically. Further, the large subterminals and main-lines have developed as the major players in the grain handling system. The 1990s have thus been the pivotal decade for the rail network on the Canadian Prairies.

Class 1 Railscapes in the Year 2000

The big four companies in the region (BNSF, UP, CN, CP) increasingly seek economy, efficiency and lower labour costs. Intermodal and coal traffics have been key growth areas. A double and sometimes triple-track state-of-the-art mainline for coal movement initially built south in the early 1980s from Gillette through the heart of the Thunder Basin National Grassland is shared by BNSF and UP (Plate 10). A million tons of coal moves out by this route every two days. The route bifurcates in the vicinity of Douglas, WY to follow either the old CNW line through Lusk or the BNSF main along the North Platte River valley. BNSF also moves coal eastwards through Newcastle, WY and Alliance, Nebraska on the old Burlington route, and on the old Northern Pacific and Milwaukee routes through Jamestown, ND and Aberdeen, SD respectively.

These modern railscapes are intrusive to say the least. The section of Thunder Basin followed by the new line is aptly named. At Guernsey, WY an air of bustle, noise and modernity at an expanded BNSF rail yard with locomotive servicing facilities on one side of the North Platte River offsets the tranquility of the Oregon Trail Ruts historic site on the other side. A visit to Scottsbluff National Monument in Nebraska is accompanied by the rumble of coal trains and the blaring horns of locomotives traversing the Scottsbluff-Gering urban area. At Lusk, WY we are warned to stay away from the railroad, for everybody's safety — a far cry from the traditional western activity of meeting the train.

Canadian Pacific's Soo Line (SOO) is gaining more and more traffic from the far west to Minneapolis — St. Paul and Chicago, and the SOO is changing in response. CP has reduced congestion in and around Moose Jaw, SK by double-tracking 20 miles from west of the city to Pasqua junction (Figure 4) at the top of the steep eastbound drag out of Moose Jaw yards. At Pasqua the Chicago-bound SOO strikes off southeast from the traditional CP main. With NAFTA, increasing numbers of trains are taking the SOO route. This increased use is also reflected in the 1998-1999 installation of new passing loop sidings at Henrik and Holloway, northwest and southeast respectively of Weyburn, and northwest of Donnybrook, ND. New grain subterminals along the SOO at Corinne and Weyburn, SK, and at Carpio, ND contribute to its visibility while some nearby branchlines have been cut back considerably.

CN's high-speed main line across the prairies from Edmonton to Saskatoon and Winnipeg connects with Chicago via CN subsidiary Duluth, Winnipeg and Pacific and the Wisconsin Central. Both CP and CN have been making efforts to gain more traffic in the enormous U.S. market as they scale down their branchline operations in Canada. CN has withdrawn from many rural areas on the Canadian Prairies, especially in southern Saskatchewan and has sold a number of lines in the northern prairies to RaiLink, a shortline company with its parent in the U.S. Both CN and CP have also profited from coal traffic originating in Alberta and British Columbia, almost all of which moves westbound to Prince Rupert, BC (CN) or Vancouver (CN and CP) for export to Japan and other Asian countries.

Huge changes have taken place in Nebraska. North Platte and Alliance in the west have emerged as high-volume rail traffic centres while in the east Omaha and to some extent Lincoln have declined. Union Pacific's main line from Chicago to the west, via its former CNW link through Iowa, now bypasses Omaha/Council Bluffs using the Blair cutoff. The UP shops and yard were removed from the cramped but valuable riverfront site at the edge of downtown Omaha (Plate 11) in the mid 1980s and replaced at spacious North Platte — a characteristic shift from

Alec Paul

Figure 4. Pasqua, Saskatchewan: new signals and double track at the western approach to the Soo Line junction, CP main 8 km east of Moose Jaw, 1996.

metropolitan real estate, valuable once vacated, to a rural location. Alliance in the northwestern part of the state has become a junction of BNSF routes for Powder River coal trains. Here the line from Gillette and Newcastle which skirts the southwest side of the Black Hills splits into routes south to Denver and the Colorado piedmont region, and east to the Missouri Valley.

Other Active Railscapes

A variety of Class 1 railway non-main lines are still operative in various parts of the northern Great Plains, along with a number of "short" and "regional" lines. Examples of the latter include Carlton Trail Railway, Central Western, RailLink, Southern Rail Cooperative on the prairies and Dakota, Missouri Valley & Western, Dakota, Minnesota & Eastern and Red River Valley &Western (RRV & W) on the northern U.S. plains (Lewis 1996). We will look at just a few of these lines here.

In North Dakota, a 1980s addition to the railscape is the Farmers Union elevator just outside Monango in the extreme south (Figure 5). Here huge piles of wheat, corn and sunflowers arose each fall in the 1980s and 1990s (Figure 6) after surrounding elevator storage was drastically reduced. Monango was a Milwaukee (railroad) town with the SOO crossing from east to west just south of it. The former SOO trackage in Monango is now run by RRV & W and still ships a lot of grain. The old Milwaukee main line which cuts across the far southwest of North Dakota (Figure 5) was the subject of intense attention in the late 1970s when the company planned to abandon it. The State of South Dakota purchased it and the line is still there, operated by BNSF whose chief interest is moving coal from southern Montana via Aberdeen, SD (Figure 7), to the upper Mississippi valley and Midwest.

In southern Saskatchewan, around and south of Regina and Moose Jaw (Figure

8), some CP and CN lines still see a fair amount of activity. Their landscapes retain some highly visible older structures to which have recently been added concrete elevators at Congress and Assiniboia. Grain trains of fifty or a hundred hopper cars travelling at moderately high speeds are still seen on these routes (Figure 9).

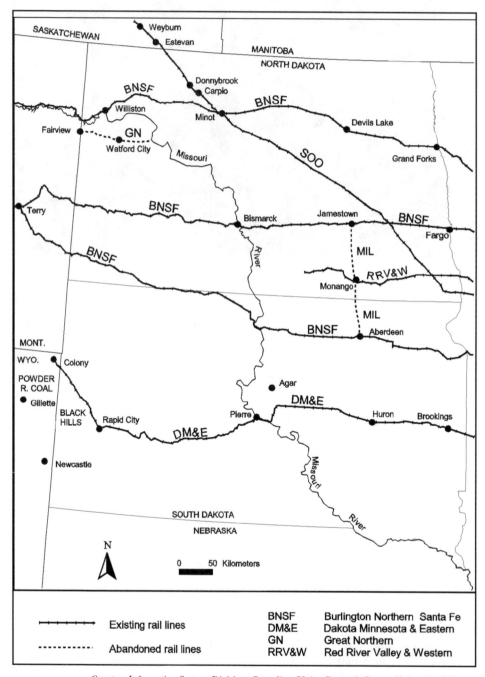

Courtesy Information Systems Division, Canadian Plains Research Center, University of Regina
Figure 5. Railways mentioned in the Dakotas.

Alec Paul

Figure 6. Monango, North Dakota: new Farmers Union elevator, RRV & W line, 21 October 1987.

Alec Paul

Figure 7. Aberdeen, South Dakota: former Milwaukee depot, BN units, State track, 14 October 1987.

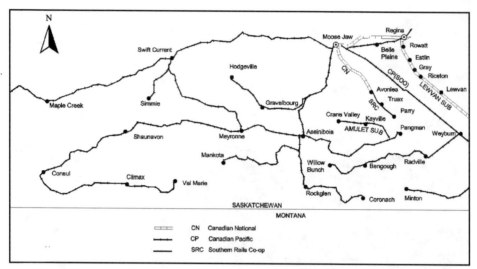

Courtesy Information Systems Division, Canadian Plains Research Center, University of Regina
Figure 8. Remaining rail lines in southwestern Saskatchewan, August 1996.

However, long stretches are almost devoid of railscape features apart from the single track and the road crossings and occasional warning signs. Mossbank, with its one elevator, is the only village remaining, along the entire 90 km of line between Moose Jaw and Congress, 10 km north of Assiniboia. While Congress has a United Grain Growers subterminal, a huge new Saskatchewan Wheat Pool terminal (Figure 10) was completed in 1999 in Assiniboia.

CN has almost withdrawn from Saskatchewan south of the CP main line. However, CN's Regina to Moose Jaw feeder is still heavily used since it has connecting spurs to a large fertilizer plant and potash mine north of Belle Plaine, SK. The CN has a small yard in northeast Moose Jaw with a wye (triangular junction) for turning trains (at the entrance to the abandoned line to Central Butte), and uses an awkward connection from the Moose Jaw CP yards to interchange cars with CP. This involves a sharp climb along Home Street in the South Hill district of the city and a subsequent reversal to cross a lengthy bridge over the CP and the Moose Jaw River (Plate 12, Figure 11). This line crossing the bridge makes an end-on junction with the Southern Rail Cooperative line (SRC, ex-CN) to Avonlea and Truax.

The grain branchlines in southern Saskatchewan have historically been served by slow-moving trains which spot small numbers of cars at the various elevators for loading when needed. In 1996, there were still some boxcar-only movements, from Simmie (Figure 12) and McMahon, on lightweight-rail tracks that would be abandoned in the fall. Today all grain moves off the branches in covered hopper cars, and surviving elevators are equipped with spouts and support equipment for top-loading these cars. Along these lines, numerous elevators have been closed and many demolished. In some cases, the branchline may still exist but the elevator-siding tracks have been removed, as at Mazenod, SK on the Mossbank-Hodgeville branch.

Shortline railscapes in the northern Great Plains have expanded dramatically south of the Forty-Ninth Parallel but have come only recently to the Canadian prairies. In extreme southern Saskatchewan, SRC was the first. It started in the late

Alec Paul

Figure 9. North of Congress, Saskatchewan: 50-car loaded grain-hopper train powered by CP 5754 and 3112 heads for main line at Moose Jaw, 24 October 1999.

Alec Paul

Figure 10. Assiniboia, Saskatchewan: new Saskatchewan Wheat Pool terminal grain elevator, 24 October 1999.

Alec Paul

Figure 11. Moose Jaw, Saskatchewan: CN grain train on Moose Jaw River bridge as CP intermodal eastbound begins climb to Pasqua junction, 6 September 1999.

Alec Paul

Figure 12. Simmie, Saskatchewan: one of the last boxcar spots, 25 July 1996.

1980s by running two branches facing abandonment, from Rockglen to Killdeer (ex-CP, operated for only one year before final closure) and from Avonlea to Parry (ex-CN). Part of this second line survives today. In 1999 a severe rainstorm washed out the trackbed northwest of Parry, and service was cut back to Truax, 15 km southeast of Avonlea. Until 1997 traffic was handed over to CN at Avonlea (Plate 13) after which time SRC took over the Moose Jaw to Avonlea section too. In 1999 SRC contracted to run the line from Assiniboia to Pangman for Red Coat Trail Railway, a separate company established to maintain rail service to remaining elevators along this former CP branch.

Railscapes of Abandonment

Minton, SK has already been mentioned. On the American side of the border, Agar in South Dakota (Figure 5, 13, 14) on a former CNW branch, is a comparable case. Abandoned railroads have met a wide variety of fates. Those that have not been subject to commercial land-use conversion can be of great value ecologically by providing refuge for native flora and fauna whose habitats largely disappeared during agricultural settlement of the region (Braband 1981). The Prairie Wildlife Centre at Webb, SK that operated for a number of years in the 1970s and 1980s, included strips of former railbed of the CPR mainline. Further, rail lines have become popular hiking trails for bird watchers and other nature enthusiasts.

More common abandoned railscapes are those stretches once traveled by trains and now under review — or already converted — for other land uses. Such conversion is faster in areas of high population density and in urban places in particular. On the northern Great Plains, however, rural railscapes of abandonment show remarkable persistence, and a range of cases is available for study.

Alec Paul

Figure 13. Agar, South Dakota: CNW boxcar branch just east of the Missouri Valley, 21 June 1982.

Alec Paul

Figure 14. Agar, South Dakota: after rail abandonment, 14 October 1990.

Alec Paul

Figure 15. Kayville, Saskatchewan: good times on the CP Amulet Sub, 17 July 1996.

Canadian National's former Lewvan line in southern Saskatchewan (Figure 8) furnishes a good illustration of a landscape of recent abandonment (1998). McDonnell (1998) and Leopard (1999) provide photographs and commentary on the railway operations just prior to closure. Two of the villages along the line, Gray and Riceton, are within the Regina commuter-shed and will survive. Gray even has an active curling and ice rink although it has not had an elevator for many years. Riceton's Wheat Pool elevator operated until the end of rail service in 1997 (Plate 14). The hamlet of Lewvan, on the other hand, effectively died with the line. By September 1999, its single remaining P & H elevator awaited the bulldozer while the Pioneer elevator was demolished in 1997 (McDonnell, 1998).

CN had the Lewvan branchline up for sale in the fall of 1999, but realistically it was never a likely candidate for shortline working. It lies only 20 km northeast of the parallel SOO main with its abundance of elevators and subterminals. The northwest end of the line at Rowatt still sees CN trains from Regina and its elevators continue to operate. The Pioneer facility at Estlin, 10 km southeast of Rowatt along the former branch, is still open too, with a semi-trailer grain truck service moving loads to Rowatt on paved roads.

Kayville on the old CP Amulet sub in south-central Saskatchewan has also fallen on hard times. Hopes had been raised by the rehabilitation of the line through town west to Crane Valley in 1996. The local hotel's 'Dodge City Saloon' hummed with enthusiasm as workers replaced ballast and rails, and trains continued to spot cars at the local Pool elevator with its steel-bin annexes (Figure 15). Yet only two years later the line was closed. In October 1999 I traced the route once again. The Ormiston sodium sulphate works were derelict and deserted and the Pool elevators at Crane Valley and Kayville no longer had a railway. At Kayville huge bundles of railroad ties awaited removal although the trackbed still retained most of its new ballast and had sound contours (Plate 15). "Six of one and half a dozen of the other," said one of the locals, "CP wanted to abandon the line, and the Pool said they wouldn't push to keep the elevators open if the line was going to go… We don't know who to blame."

Abandonment and/or abundance? Since the death of the Crow Rate in Canada, abandonment railscapes are increasingly frequent, as newly closed lines join those from the 1980s and earlier 1990s. The case of the CP Dunelm branchline received plenty of publicity in southern Saskatchewan. This was a 40-km boxcar branch from Player (southwest of Swift Current) to Simmie, which was abandoned in the fall of 1996. In July of that year I was able to photograph one of Simmie's last "boxcar spots" (Figure 12). Other pictures of the final weeks of the line are available in McDonnell (1997, 1998). Just prior to abandonment, the trackbed was in poor shape and trains crawled along at 15–20 km per hour. A year later I had the opportunity to walk a short stretch of the line after closure. Everything was still there but plants were three or four feet tall and the rails were often hidden. I hopped along the ties, thinking about what had proved a futile effort on the part of the local committee to save the branch for shortline working (Paul and Bratvold 1992). I found the hike neither easy nor agreeable.

Goodman and Reinhold (1993) describe the landscapes of one of the old Great Northern lines across North Dakota. This stretch of line has been abandoned for a number of years yet it remains very conspicuous feature on the plains. An enormous embankment filling in a former trestle still traverses a coulee a short distance

Alec Paul

Figure 16. West of Watford City, North Dakota: the massive GN enbankment near Arnegard, 25 September 1996.

west of Watford City — characteristic of main line rather than branch engineering (Figure 16). The section of the line from Watford City west to Fairview, Montana (Figure 5) was still operated as a grain branch until the early 1980s and as of 1999 a few elevators still stood along it. Further east beyond Watford city, a fine bridge across the Missouri River once accessed by the state's only railway tunnel is still in existence; though long abandoned by the railroad, it now makes an attractive recreation landscape.

Current Trends and a Prospective to 2010

The changing railscapes illustrated here are symptomatic of the evolving geography of North American railways. Remaining main lines on the northern Great Plains compete for a long-distance traffic which has both a traditional east-west axis and a rapidly expanding north-south dimension. The two Canadian Class 1 lines have embraced this second dimension as they adjust to an ever-more continental economy and move increasing volumes via Chicago. Containers from the Pacific Rim to Toronto may be routed via any one of several west-coast ports and a variety of rail routes. The all-Canadian routes via Vancouver and rails north of Lake Superior are a diminishing option in this day and age. Their greatest ally may well be the rail-traffic and congestion problem in and around Chicago.

Traffic on the main lines of this region must be interpreted from a continental perspective, and their changing railscapes understood in the same fashion. Probably the main lines in 2010 will be much the same as today, with an increase of traffic on the SOO and ever more coal being moved out of Powder River fields on BNSF and UP.

The railscapes away from the main lines will be highly volatile. The Class 1 carriers will sell off more feeders and close more branches. Shortlines and regionals will be under growing pressure to perform larger grain handling roles. Some will be unable to respond and will disappear, while others will upgrade and expand. There is a clear trend to more concentrated ownership of these lines as their growth in capital requirements will continue. North of the border, there will be an increasing awareness of the pervasive American role in ownership of shortlines and regional lines.

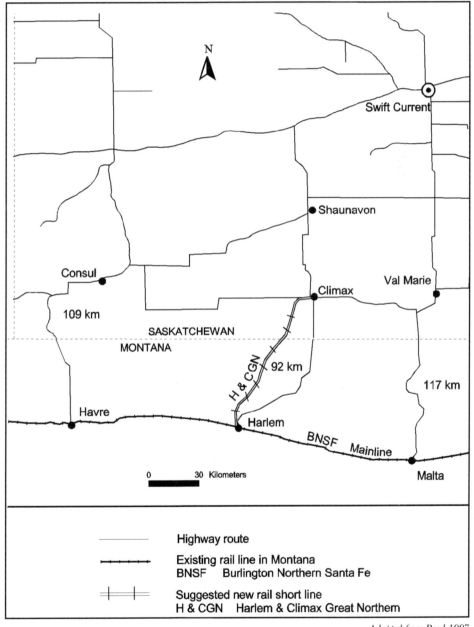

Adapted from Paul 1997

Figure 17. Suggestion for an international short line.

Fire and flood will continue to play a part in the landscape changes on the short-lines. Fires will continue to claim wooden grain elevators, railroad trestles and bridges. Flooding will wash out trackbeds and undermine bridge approaches. In some cases these events will provoke abandonments, especially of lightly-used sections which probably have only a few years of life left anyway. They will also contribute to the difficulty of converting abandoned trackbeds to use as recreational trails.

South of the border especially, regional railroads may pursue a larger vision of growth which could even entail some new construction. The most publicised case is the DME. Its proposed expansion would require it to by-pass some cities through which it now runs, notably Pierre, the South Dakota state capital, as well as building up to 280 miles of totally new route.

Changes will continue into the future. For example, the likely elimination of Canadian rail service to far southwestern Saskatchewan could lead to a situation reminiscent of that in the Kootenays of British Columbia in the late nineteenth century. Canadian railways had not yet been built in this region during the 1890s, so an American railroad, the Great Northern, jumped in to fill a void. Today the void is further east, but let us hypothesize a new Great Northern, reborn as an international shortline to Climax from Harlem, Montana (Figure 17, from Paul 1997). A new subterminal at Climax would be the northern end. In true NAFTA style, Canadian grain exported this way via BNSF could find an outlet to world markets via Portland, Oregon. Whether DMV & W or Montana Rail Link (should it take over BNSF's northeastern Montana branches) might be prepared to build across into southern Saskatchewan, along the lines of this proposal of the "Harlem and Climax Great Northern," is another question.

Conclusion

Railways have had profound influences on the landscapes of the northern U.S. plains and Canadian prairies, both directly in the sense of creating their own railscapes, and indirectly through their promotion of settlement and the growth of a commercial, largely agricultural economy. This paper has concerned itself mostly with railscapes which, at the start of the new century, are in a state of flux. Substantial changes have taken place since the 1960s and 1970s. Even in 1995 there were still a few boxcar grain branchlines on the Canadian prairies but these are now history. South of the border, grain traffic is still very important, but coal traffic from Wyoming and to a lesser extent Montana has had a huge impact on the main lines. Increased long-distance intermodal traffic which demands speed and reliability has provoked large investments even on those main lines not significantly impacted by coal traffic.

The railscapes of many of the old branch lines have become landscapes of abandonment. Where such lines are still being operated, mostly as shortlines, they may have been upgraded, with some new grain subterminal facilities built along them. In other cases, trains continue to serve older elevators on tracks which suffer from inadequate maintenance. These latter stretches are the next candidates for closure. Perhaps the current concerns over impacts of such closures on rural roads and rural lifestyles in general will help to change the thinking about the role of railways in the northern Great Plains. Hopefully railways can continue to provide a valuable service within the region besides their expanding involvement in simply moving long-distance traffics through it.

Acknowledgements

Thanks to Teri Rogoschewsky for her patient word processing, to Diane Bolingbroke for her work on the figures, and to Todd Radenbaugh for his constant encouragement and helpful criticism. Discussions with Don Berg, Geography, South Dakota State University and John Everitt, Geography, Brandon University are gratefully acknowledged. I also benefited from comments by participants at several conferences where earlier, shorter versions of this evolving work were presented. But mostly this paper resulted from fieldwork observation, during the course of which conversations with numerous local people and anonymous current and former railway employees proved most enlightening. Remaining shortcomings are of course my responsibility.

CHAPTER 9

PLAINS CLIMATE CONDITIONS: AN EARLY INTERPRETATION BY THE SMITHSONIAN METEOROLOGICAL PROJECT IN MANHATTAN, KANSAS 1858–1873

Karen DeBres

"The observer ought, however, in all cases, to render his results as accurate as possible, and not to consider them of little value even though they should fall below his ideal standard of perfection."

Instructions for Smithsonian Voluntary Observers, 1858

"Kansas brags on its thunder and lightning and the boast is well founded. I never before observed a display of celestial pyrotechny so protracted, incessant, and vivid as that of last Saturday night."

Horace Greeley in Lawrence, Kansas, May 1859

ABSTRACT. The Smithsonian Meteorological Project's records are a rich archive for the study of the science and culture of mid-nineteenth century America. This essay concerns the first attempt to re-image the Great Plains climate on the basis of daily weather data. Nineteenth-century theories and claims about Great Plains climate changes are reviewed first, followed by a brief discussion of the Smithsonian's meteorological "Project." Observations by the three volunteer observers in Manhattan, Kansas are then summarized. Various weather events of the period (droughts, tornadoes, hailstorms, and floods) are discussed as well as the 1867 earthquake. The wide daily temperature fluctuations of the Plains climate contradict the immigrant guidebooks, which had contained (probably for reasons of both economic and political expediency) overly optimistic claims of temperate weather conditions.

Isaiah Bowman, a distinguished early twentieth-century geographer, once said that facts more valuable than all the gold in the Klondike are contained in the old records of the Weather Bureau (Mohler 1948). Included in these records are the files of the Smithsonian Meteorological Project, which predate the establishment of the Weather Bureau itself. From 1849 to 1874 between 300 and 500 volunteer observers across the country sent monthly weather logs and related information to Washington, DC. These records are a time capsule of weather conditions and related environmental material kept on a daily basis, and provide a rich archive for the study of the science and culture of mid-nineteenth century America. They are especially valuable to scholars interested in the history of science in the Great

Plains, since they often contain the first daily environmental data from the region's frontier and early settlement period.

This article concerns the first attempt to re-image the Great Plains climate on the basis of daily weather data. The climate of the Great Plains region, extending from Texas and Oklahoma well into the prairie provinces of Canada, was the subject of conjecture, speculation and hyperbole during the nineteenth century. Two opposing images proposed to describe the region, "garden" or "desert," are almost banalities today, but during the Plains settlement period such images were crucial.

Some nineteenth century explorers posited the existence of an authentic desert east of the Rockies (Bowden 1976) while railroad companies and town speculators, intent on luring farmers west, described the Kansas portion of the Great Plains at least, as the "garden of the world" (Travis 1978). One outcome of the Smithsonian Project was to publicize the daily weather and related environmental conditions of the Great Plains (as well as the rest of the United States). By doing so, the weather differences in the Plains and the weather in the better known eastern regions could be more accurately compared.

Beginning in 1858, the Manhattan weather station became part of the network established by the Smithsonian Institution. Manhattan is located about 110 miles west of Kansas City and it was the most "westward" of the Kansas Territory stations.

The discussion to follow is divided into four sections. In the first, theories and claims prevalent in the mid-nineteenth century about changes in the Great Plains climate are briefly reviewed. The second section introduces the use of voluntary weather observers in the United States and then describes the Smithsonian Meteorological Project in the context of the newly established Smithsonian Institution. In the third section, "The Project," Manhattan weather conditions and related observations, as well as the reactions of the eastern-educated scientists in Manhattan recording the observations, are also described. Information provided by the Smithsonian observers is then related back to some of the contemporary theories of climate change on the Plains.

Nineteenth-Century Myths of Great Plains Climates and Climatic Change

Kansas is situated in the geographic center of the forty-eight contiguous states. It is located 37° to 40° north of the equator and is subject to a continental climate (i.e., extreme temperatures and variable precipitation). The record high temperature is 121°F and the record low is -40°F. Surface winds bring moisture to Kansas from the Gulf of Mexico mostly during the summer and average annual rainfall amounts vary from forty inches in southeast Kansas to about fifteen inches near the Colorado border (Self 1978; Eagleman and Simmons 1985). The wide range of weather conditions found in the Great Plains results from the distance from any major body of water and from the presence of different air masses that frequently alternate in their dominance of the region (Rosenberg 1986: 23). The Smithsonian observer data reveal the resulting continentality in both the annual precipitation for Manhattan, 1855–75 (Figure 1) and monthly and annual mean temperatures for Manhattan 1858–73 (Table 1).

Reports by early nineteenth-century explorers encouraged a belief that the Plains could never become an agrarian region. The heat of the high summer and the lack of water were duly noted. Zebulon Pike predicted in 1806 that "these vast plains of the western hemisphere, may become in time equally celebrated as the sandy deserts of Africa" (Meinig 1993: 76). Edwin James, the geographer of the official expedition of 1819–20, reported:

> In regard to this extensive section of the country I do not hesitate in giving the opinion, that it is almost wholly unfit for cultivation and of course uninhabitable by a people depending upon agriculture for their subsistence. (Meinig 1993: 76)

The historical geographer Ralph Brown credits the railway surveys of 1855 with the first challenge to these opinions. By 1870 Ferdinand Hayden of the Geological Survey said that "every year as we know more and more about the country this (desert) belt becomes narrower and narrower, and as a continuous area has already

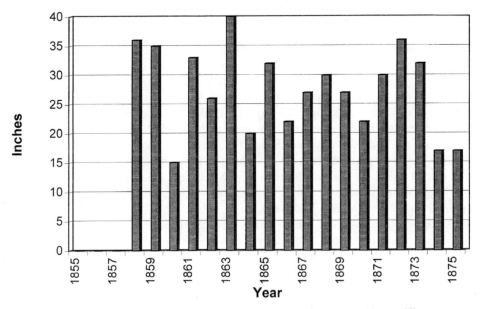

Figure 1. Variation in Annual Precipitation, Manhattan, KS, 1855–75. Source: *Climate of Kansas.*

Table 1. Monthly and Annual Mean Temperatures for Manhattan, Kansas, 1858–73

Year	Jan.	Feb.	Mar.	Apr.	May	June	July	Aug.	Sept.	Oct.	Nov.	Dec.	Annual
1858		25.5	46.1	51.4	60.0	75.2	80.8	74.6	69.7		33.7	25.8	
1859	32.0	33.9	46.9	50.6	64.4	75.1	82.2	78.0	66.9	54.0	46.0	21.8	54.9
1860	23.1	32.6	52.4	56.2	65.7	81.8	83.9	85.0	72.6	55.7	37.8	33.4	57.1
1861	24.9	35.9	42.6	56.3	64.9	79.1	78.8	78.8	69.0	56.6	41.4	32.5	55.1
1862	18.2	24.1	38.6	50.8	69.0	74.1	80.6	70.8	69.2	57.6	44.1	38.6	53.2
1863	36.1	30.1	47.0	61.2	79.4	70.9	76.1	78.6	73.2	48.3	38.1	25.7	55.4
1864	24.0	32.2	38.3	48.3	64.8	75.8	82.0	78.3	70.8	46.0	36.6	28.0	52.1
1865	27.4	35.2	42.0	54.7	68.2	76.1		75.4					
1866					61.8	77.8	80.8	72.7	62.3	57.5	46.4	31.0	
1867	22.4	32.3	24.8	50.0	59.1	73.9	75.7	77.8	69.9	58.9	45.1	35.9	52.2
1868	18.4	27.7	49.4	46.6	67.1	74.3	83.7	76.8	60.7	53.1	38.1	25.3	51.8
1869	30.8	31.3	35.7	47.4	58.0	66.4	73.5	76.0	61.9	44.8	37.3	30.2	49.4
1870	27.7	34.1	36.2	53.5	65.8	74.6	84.9	72.9	68.3	56.6	45.6	30.1	54.2
1871	28.8	36.0	47.2	60.0	67.2	77.5	79.4	76.2	67.2	57.6	37.0	24.7	54.9
1872	25.2	32.9	38.7	56.6	65.7	75.0	78.7	77.7	66.2	52.4	34.1	21.1	52.0
1873	19.7	30.7	42.5	47.8	62.8	76.6	77.4	78.8	66.4	51.9	42.1	29.1	52.1

Source: Kansas State Board of Agriculture.

ceased to exist even in the imagination" (Brown 1948: 371). But the period between the 1850s and the 1870s was crucial to parts of the Plain's settlement, particularly in eastern Kansas. Walter Prescott Webb believed that the 1850s decade was the high point of the general acceptance of the "desert" idea, but its popularity lasted until the 1870s — long enough to stop the waves of the frontier but not enough to stem the tidal surge of population after the Civil War (Bowden 1976: 134). Given the presence of the "desert" belief, it is not surprising that those intent on settling this region looked for ways to make the climate seem more appealing, especially by suggesting that rainfall amounts were high and that the climate was more conducive to settlement than it had originally been reported.

Moreover, several versions of a changing climate theory became common in the mid-nineteenth century (for related discussions of this topic see Brown 1948; Blouet and Lawson 1975; Meinig 1993; Olson 1995). These often suggested the same outcome: the Plains climate was dramatically improving and becoming more like the humid East, with more precipitation and fewer extreme weather events such as tornadoes and droughts. One theory was that planting trees in the Plains grasslands would cause rainfall to increase; this was Ferdinand Hayden's opinion, who wrote in 1869 that tree growth would improve the climate, "which has already changed for the better along the Missouri in Nebraska" (Brown 1948: 376). To encourage this process, Congress enacted the Timber Culture Act in 1869: this law gave 160 acres of land to anyone who would plant part of it in trees. The bill was introduced by Senator P.W. Hitchcock of Nebraska, a neighboring Plains state. Hitchcock called the law "a major step toward solving the problem of the dry climate of the Plains" (Manley 1993: 17).

A related climate change theory, often referred to as "rain follows the plow" was also popular, arguing that the breaking of the prairie sod and the planting of row crops as well as trees would increase the amount of precipitation on the Plains. As early as 1873 Frank Snow, Professor of Natural Sciences and Meteorology at the University of Kansas, said that it was impossible to present evidence for this without fifty years of data, despite what was seen as an increase in rain since the early settlement period (Snow 1873). But writing in the same year in an immigrant tract entitled *The Homestead Guide*, F.G. Adams claimed that "the fall of moisture on the plains is steadily on the increase as cultivation and tree-growing advance, and as the Indian, the buffalo, and the prairie fires cease to prevail" (Adams 1873). Another climatic change theory held that building railroad and telegraph lines increased rainfall because of "the effect of the electrical currents running on rails and wires. In the case of the railroads the effect is more probably produced by the disturbance of the atmosphere attendant upon the rushing of the trains" (Manley 1993: 18).

That the climate of the Great Plains was conducive to agriculture was widely publicized in immigrant guides of the 1850s through the 1880s. In 1870 the author of *Kansas as She is, The Greatest Fruit, Stock, and Grain Country in the World* claimed:

> The climate of Kansas is, without exception, the most desirable in the United States, it is better than that even of the same latitude, east of the Mississippi River. ... Since the year 1860, the State has been blessed with an abundance of rain... The oldest inhabitants universally agree that the drouth of 1860 was the only one of any consequence that ever visited Kansas. (Kansas Publishing Company, 1870: 8)

The Kansas climate was praised not just by railroad companies and immigrant guides in general, but also in a publication designed to encourage "Free Staters" into Kansas. *Information for Kansas Immigrants* was produced in 1856 by the New England Immigrant Aid Society. It described the Kansas climate as one with less variation than that of southern New England, stating "we believe, as a general rule, the variations there will be less frequent and extreme than they are liable to be in this section of the country" (Webb 1856: 12).

The activities of the New England Immigrant Aid Society were closely aligned to those of the new Republican Party. The Party's 1856 campaign chant "Free Soil, Free Labor, Free Men" was often applied to land in the American Great Plains, and particularly to the newly opened Kansas Territory. The Republican Party supported free state expansion (as opposed to slave state expansion) into arable western lands, suitable for yeoman farmers and their families. So in the 1850s, denial of the "desert" image of Kansas could be considered a function of one's allegiance to the Republican Party (Emmons 1975: 128). Environmental images, it can be argued, had been altered just before the establishment of the Smithsonian Project in Kansas — not by science, but by politics.

Early Weather Records by Voluntary Observers

In the United States, voluntary weather observers have been indispensable for meteorological research since the seventeenth century (see Fleming 1990: 81–83). Writing in the nineteenth century, Sir John Herschel said of meteorological research that it "can only be effectually improved by the united observations of great numbers widely dispersed ... (it is) one of the most complicated but important branches of science ... (and) at the same time one in which any person who will attend to plain rules, and bestow the necessary degree of attention, may do effectual service" (Fleming 1990: 81). In 1948 the State Meteorologist for Kansas said of voluntary meteorological observers that no other group of people in the country gives so unselfishly of their time, without monetary compensation, in making daily observations over long periods of years for the benefit both of their communities and the nation (Mohler 1948).

Josiah Meigs, Commissioner of the General Land Office, first tried to collect weather observations on a national scale in 1817. Two years later, the Surgeon General's office began keeping records at military posts. Each hospital surgeon was ordered to keep a diary of the weather, and attempts were made to correlate local weather conditions to the health of the men under his care. Data were collected at eighteen forts in 1819, at sixty-two by 1842, and at 159 by 1855. Daily temperatures were consolidated into figures of mean daily, mean monthly and mean yearly temperatures (Lawson 1974: 34–35). In 1837, James Espy, the first official U.S. government meteorologist, proposed that a network of three hundred observers be established across the country. Several ideas based on these efforts became part of the Smithsonian Meteorological Project.

The Smithsonian Institution and the Smithsonian Meteorological Project

The Smithsonian Institution was established on the basis of a bequest from James Smithson, a wealthy Englishman who was also an amateur scientist. In 1835, Smithson left more than half a million dollars to "found at Washington ... an establishment for the increase and diffusion of knowledge among men." Congress accepted the bequest in 1836. Joseph Henry, the leading American physicist of the day, was selected as chief administrator. By 1859 the Smithsonian was already promoting studies in the fields of astronomy, geography, geology, and philology. In meteorology it

> has established a system of meteorology consisting of a corps of several hundred intelligent observers, who are daily noting the phases of weather in every part of the continent of North America. It has imported standard instruments, constructed hundreds of compared thermometers, barometers, and psychrometers, and has furnished improved tables and directions for observing, with their instruments, the various changes of the atmosphere, as to temperature, pressure, moisture, etc. It has collected, and is collecting, from its observers, an extended series of facts, which are yielding deductions of great interest in regard to the climate of the country, and the meteorology of the globe. (Rhees 1859: 8)

This program was entitled the Smithsonian Meteorological Project and was the first research project funded by the Smithsonian. Its purpose was to chart the paths of storms, to explore the climate of North America, and to become the national center for atmospheric research. The Project's centerpiece was an extensive system of hundreds of volunteers who kept weather logs on a common monthly form. Long-range data on climate conditions over a series of years were assembled from stations across the country, providing a sound foundation for the study of meteorology in the United States. At one point the system included six hundred volunteers. This program gave many American citizens a grass roots connection with their new scientific organization. Henry described the system in December 1848, in the *Second Annual Report of the Secretary to the Board of Regents*:

> An appropriation of one thousand dollars was made at the last meeting of the Board, for the commencement of a series of meteorological observations, particularly with references to the phenomena of American storms.
>
> It is contemplated to establish three classes of observers among those who are disposed to join in this enterprise. One class, without instruments, to observe the face of the sky as to its clearness, the extent of cloud, the direction and force of wind, the beginning and ending of rain, snow, and etc. A second class, furnished with thermometers, who besides making the observations above mentioned, will record variations of temperature. The third class, furnished with full sets of instruments, to observe all the elements at present deemed important in the science of meteorology, it is proposed to employ, as far as our funds will permit, the magnetic telegraph in the investigation of atmospherical phenomena. The advantage to agriculture and commerce to be derived from a knowledge of the approach of a storm by means of the telegraph, has been frequently referred to of late in the public journals. (Henry 1886: 285–86)

The observers were asked to fill in three large double-page sheets a month, tabulating and recording their data, and to maintain their instruments in good repair. Each observer was issued twenty-four of the monthly forms a year, and was asked to return one set to the Smithsonian. Required data collection was written down on the first sheet, with categories entitled "for use of the barometer, hygrometer, to note rain or snow, the presence of clouds, the direction of the winds, force or pressure vapor in inches, and relative humidity or fraction of saturation." There was a brief "explanation of the above columns" at the end of the page. The second double page was entirely given over to the description of casual phenomena. This information in particular could be a valuable resource for students interested in local weather conditions or in the observation of phenomena such as aurora borealis, meteorites, or the changes of the seasons. Observers were asked to comment upon:

> Thunderstorms — time of occurrence and direction of motion. Tornadoes, lightning, aurora borealis, shooting stars, solar and lunar haloes, parhelia and paraselenes, frost, depth of ground frozen, closing and opening of rivers, lakes, canals and streams, temperatures of wells and springs, earthquakes. (National Weather Service Archives)

The importance of thunderstorm observation is underscored by its appearing first on the list of instructions. The third double page consisted of general remarks and directions devoted to using the instruments. From 1858 until its termination in 1874, this Project averaged 20 percent of the Institution's budget in publications and research (Fleming 1990: 76). The data collected were made available quickly, with local newspapers often publishing accounts based upon volunteer observations. At its inception, a daily weather map compiled from telegraphic reports received at ten o-clock every morning was on display at the Smithsonian's building in Washington DC. Between 1852 and 1884, Arnold Guyot, the Princeton scientist who both selected the instruments and determined the observation times, produced four editions of the *Collection of Meteorological Tables.*

Meanwhile two other nationwide meteorological systems were established in the 1860s and 1870s, which eventually helped to end the Smithsonian Project. The first was a program of monthly weather bulletins and crop conditions, begun by the Department of Agriculture in 1863, which was also based on information gathered by voluntary observers. The second was the Army Signal Service which in 1870 began to collect and to tabulate weather data, and became the forerunner of the National Weather Service. All three of these systems operated in Kansas.

The Project in Manhattan, Kansas 1858–1873

Isaac Tichenor Goodnow (1814–94) was the first volunteer observer of the Smithsonian Meteorological Project for Manhattan, Kansas. He recorded most of the observations between February, 1858 and August, 1863. Goodnow was Professor of Natural Sciences and Languages at the Providence Seminary in Providence, Rhode Island, when he and his brother-in-law, the Reverend Joseph Denison, decided to bring their families west as "Free Staters" in 1855. Both Goodnow and Denison were instrumental in founding the Methodist-supported Bluemont Central College, a preparatory school, in Manhattan. Part of Bluemont College's focus was upon agriculture, and its agricultural fields and other assets were donated to the Kansas State Agricultural College when the latter was established in 1863.

Despite the claims of the immigrant guides, the new arrivals to the Kansas Territory would have found the continental climate full of typical temperature extremes. The first winter experienced by many of the new settlers in 1855–56 was one of the most severe: Kansas historian Kenneth Davis describes a winter of deep snows, high winds, and sub-zero temperatures which resulted in the death of both settlers and animals (Davis 1977: 74). Between December, 1855 and March, 1856 the Kansas River, along which many of the new settlements were located, was "bridged with ice" (Dary 1987: 185). The winter of 1858–59 was also very cold. Table 1 shows a monthly mean temperature for December 1858 of 25°F. Goodnow described December as:

> A very cold month ... a north wind with not a day to thaw so as to produce
> mud one or two weeks. The Big Blue and the Kansas rivers have frozen
> over, ice eight inches thick. Four days since multitudes of cat and buffalo
> fish were speared through holes in the ice to which they flocked to get air.

On February 14, 1858, Goodnow recorded in his diary that it was "the coldest
day of the season -19-16-4- ([temperatures] at 7:00am 2:00pm and 9:00pm).
Remained at home, read and wrote ... dreary. Thankful I am not obliged to be out
in the snow." That week Goodnow also finished making a weather vane for use in
his observations.

Although the first winters the settlers experienced may have been very cold, the
summers were generally very hot. The summer of 1858 was especially so, with
Goodnow recording four days in June over 100°F, and one day in July at 110°F. The
heat continued into September, with temperatures well into the nineties in both
August and September, and another daily high temperature of 100°F in August.
The summer of 1860 was also very hot, with one day in August reaching a high temperature of 112°.

Unfortunately, the records from January to June 1859 are missing, which
included a tornado in Manhattan. During a visit, however, Horace Greeley, the
well-known newspaper editor and journalist, described Manhattan as "an embryo
city of perhaps one hundred houses, of which several were unroofed and three or
four utterly destroyed by a tornado on the wild night I passed in Atchinson (15th
inst. [May])" (Greeley 1860: 46).

Most of the record for 1860 is also incomplete as Goodnow was evidently not in
Manhattan for the first half of the year, leaving the Smithsonian Project records in
the hands of the Reverend N.B. Preston, who used the wrong form. Preston does
note the great heat of that spring, recording for example 95°F on April 26, and 97°
on May 24. Goodnow returned to Manhattan in time to describe a tornado on July
31, observing the importance of wind direction:

> Beginning at 5:30am a strong wind blowing from the S at the time. Storm
> cloud came from the N attended by a tornado unroofing buildings.
> Lightning forked and zigzag. Ended at 10pm.

Goodnow recorded the temperatures for that day as 82°F at 7am, 100°F at 2pm, and
65°F at 9pm.

The record from Manhattan, as well as that of Leavenworth and Larned (an
army post further west), indicates that the drought of the early and middle 1860s
ranks with that of the 1930s as one of the most severe ever observed in Kansas
(Mohler 1948: 123–24, and Figure 1). The drought lasted roughly from October
1859 until March 1868 in the settled parts of Kansas, 1860, 1861, 1862, 1864, and
1867 being the driest years. Annual precipitation in Manhattan was 4.7 inches
below the normal average precipitation of 30 inches during the 1860 to 1864 period (Mohler, 1948). Most of the settlers had arrived from the more humid regions
to the east and were unaware that this would be the worst drought in Kansas for
seventy years. The 1860s drought is believed to be responsible for the adoption on
the state seal of the motto "Ad Astra Per Aspera" (To the Stars Through
Difficulties) (Mohler 1948: 123).

On September 13, 1860, Goodnow records the first instance of an aurora borealis. These phenomena, although almost unknown today, were fairly common in
mid-nineteenth century Kansas. Among others, Bone has noted that there were

very large numbers of sunspots in 1870. The most extensive auroras tended to be those in the pre-sunspot maxima phase (Bone 1991: 31).

Throughout much of the Great Plains hailstorms are common events (Changnon 1977); Goodnow recorded one instance of very large hailstones on June 17, 1862:

> Some hail stones 2 x 1 inch. Several panes of glass broken out of College private dwellings. Storm from the nw-forked lightning. Several fields of wheat up Wildcat (Creek) were ruined — hens killed by hail stones.

In August of 1863 Goodnow transferred the work of the Smithsonian Meteorological Project to his nephew, Henry Denison; a member of the Kansas State Agricultural College class of 1867, Denison was the official voluntary observer from September 1863 to September 1865. Unlike Goodnow, his "Casual Observations" are not concerned with phenomena appearing in the skies, but with the flora and fauna, especially birds. In July of 1864 Denison records finding a nest of red wing blackbirds, for instance, and mentions seeing or hearing swallows, prairie chickens, and meadowlarks. He records the date for one of the harbingers of spring in this region, the flowering of which he calls the "violet red bud hornbeam", more commonly called the red bud, on April 20, 1865. Red buds are part of the tree canopy found in the wooded areas in eastern Kansas, and are common in many gardens.

Table 2. Weather Events and Other Notable Phenomena Recorded in Manhattan, Kansas, 1858–73

May 8, 1858	aurora borealis	Goodnow	"Pillars of light with capitals like the Corinthian order"
August 26, 1858	high wind	Goodnow	"Several frame houses blown down"
December 28, 1858	ice	Goodnow	"Big Blue broke up-running ice prevented ferry boat from crossing"
August 9, 1859	meteor	Goodnow	"Meteor in the N. at 10pm appeared the size of a hen's egg"
July 31, 1860	tornado	Goodnow	"Tornado unroofing buildings, lightning – Forked and zig-zag"
August, 1860	high temperatures	Goodnow	"112 F."
January, 1861	snow	Goodnow	"18 inches … a severe storm of snow attended by High wind"
January 15, 1862	solar halo	Goodnow	"Solar halo with two bright spots … nearly as large and bright as the sun"
July 1, 1864	flood	Denison	"Kansas river rose six feet, current very Swift"
March, 1865	cold	Denison	"Kansas river frozen over so that men cross on the ice"
February, 1867	flood	Mudge	"Streams higher than known since the settlement"
April 24, 1867	earthquake	Mudge	"The college building was violently shaken. The clocks were stopped"
August 1867	grasshoppers	Mudge	"Grasshoppers … come on the 7th and depart on the 11th"
March, 1869	zodiacal light	Mudge	"Seen in the evening; after sunset … as brilliant as the milky way"
July 1870	heat	Mudge	"The hottest day for ten years"
March 6, 1871	spring	Mudge	"White maple blossomed, robins appeared"
January 28, 1873	cold	Mudge	"14 degrees – coldest day in 13 years"

Between October 1865 and May 1866, there is a break in the record, until it is taken up by Denison's successor, Benjamin Franklin Mudge. Mudge (1817–79) was an 1840 graduate of Wesleyan University in Middletown, Connecticut and a friend of Goodnow. He had also come to Kansas as a Free Stater. Helped by Goodnow to his appointment as the first state geologist and becoming the holder, in 1866, of the Chair of Natural Sciences and Higher Mathematics at the Kansas State Agricultural College, he was the obvious choice to continue the role of voluntary observer for the Smithsonian.

Daily weather records had been kept in Manhattan for eight years when Mudge took over as the official voluntary observer in May 1866. With Mudge there is a change, with more references to climate extremes in the "Casual Observations" section (see Table 2). Mudge rejected the claims of the immigrant guides, which often said that the Kansas climate was either "improving" quickly from the extremes noted by the early explorers or was of a more temperate nature than those climates of the eastern states. Indeed, Mudge's interests appeared to lie *with* weather extremes. In February 1867, for example, he writes, "from the 15th to the 20th the streams were higher than known since the settlement of the state." On February 12, this flood on the Kansas washed out the pontoon bridge that crossed the river at Topeka. In the next month Mudge is impressed by the cold weather, saying, "this has been the coldest March known to the oldest settlers." He recorded an earthquake on April 24, 1867:

> At 2:32pm a shock of an earthquake was felt lasting ten seconds. The shock appeared to come from the S or SW, as the water of the Kansas River was rolled in a wave at least two feet high from the South to the North bank. It appeared to be one continuous shock, though at about the third or fourth second a much heavier impulse was felt which after two or three seconds became weaker. The college building (three stories high of stone) was violently shaken but no damage resulted. The clocks were stopped, as were most in town. Some houses (stone) in this vicinity with weak walls were fractured, but none thrown down. Dishes were in some cases jarred from shelves. Cattle showed signs of alarm. Telegraphic and magnetic instruments were not affected. Most of the population badly frightened!

An important non-weather event recorded by Mudge in 1867 concerned one of the periodic grasshopper invasions that then plagued in the Great Plains, with the earliest report of a grasshopper invasion in Kansas Territory occurring in August 1854 in Neosho County. In both 1860 and 1861 crop destruction by grasshoppers was severe and associated with the first years of the drought. The average annual precipitation by 1867 (at the various observing stations in the state) was only 22 inches (Mohler 1948; Self 1978). In August 1867 Mudge mentions grasshoppers or "red leg'd locusts" in his monthly log, which he says, "come on the 7th and depart the 11th." According to the *Annals of Kansas*, the grasshoppers were a concern throughout the settled region of the state that summer (Wilder 1886); that volume notes cautiously on May 15 "grasshoppers industrious," and then discusses their departure from various counties and their flight "back to Colorado" (Wilder 1886: 457). But as late as September they are reported flying southwest out of Leavenworth in large numbers (Wilder 1886:461). These grasshopper invasions continued for years to come, with the most damaging grasshopper infestation in Kansas taking place in 1874.

Although the drought continued in 1867 with a dry summer and fall, the spring had been wet. This presented a green and prosperous Manhattan and its environs to Josiah Copley, a journalist touring Kansas on the new Union Pacific Railroad. Copley described the locality as:

> such a combination of the grand and the beautiful — the soft green of the gracefully undulating prairies, the dark rich foliage of the trees which skirted and marked the winding course of three streams, the Kansas, the Blue, and the Wild Cat, for miles north and west and east, the beautiful farms with which the rich landscape was dotted all over, with their comfortable-looking and really pretty stone houses, and the bright and quiet-looking town in the valley beneath, with the College on the rising ground in the rear is rare indeed! (Copley 1867: 52)

However, Mudge continued to note extremes in weather events. He recorded, for example, the coldest day ever known at the college on December 11, 1868, when his thermometer read -16°F. He observed the disastrous flooding of the summer of 1869 which occurred along the whole Kansas River watershed, just before and after the corn harvest. The Kansas newspapers, whose editors generally acted as "boosters" for their new state, were evasive regarding its extent due to the extensive damage to crops (Malin 1990: 13).

Despite the flooding, 1869 had a good corn harvest for most Kansas farmers. In the late 1860s Kansas still had a national reputation for droughts, however, so Henry Worrall's painting entitled "Drouthy Kansas" made light of that reputation when it was reproduced in the *Kansas Farmer* (Figure 2) (Wilder 1886): it showed happy farmers hauling gigantic fruits and vegetables to market with a rainstorm over a flooding river in the background.

Temperatures in Manhattan, meanwhile, were even more varied in the 1870s

Figure 2. "Drouthy Kansas" (Source: Zornow 1957: 244).

than they had been in the 1860s, according to Mudge. He recorded July 1870 as the hottest month since records had been kept, with twenty-six days of highs of 90°F or higher. In Lawrence, Frank Snow in the same summer noted forty-six days with temperatures exceeding 90° (Snow 1873: 41). Winters could also have extreme temperatures: Mudge recorded January 28, 1873, at -14°F. The last entry by Mudge was August 1873.

To what extent were Goodnow, Denison, and Mudge typical of the Smithsonian's weather observers? According to a study of Smithsonian voluntary observers (Fleming 1990), Goodnow and Mudge especially are excellent examples of the first generation. Fleming sampled observers for the years 1851, 1860, and 1870, and concluded that the status of observer declined from 1851 to 1870; he attributed this to the Project initially attracting a good deal of attention from the American scientific community and to the changing nature of that community. Observer professors of 1851 were replaced at their institutions by "individuals who typically had more formal education, more professional opportunities for research and publication in their specialties and less time to spend as passive and poorly rewarded weather observers" (Fleming 1990: 93). The generalist professors were succeeded by a younger generation of specialists, who called themselves biologists or zoologists or geologists rather than natural scientists. This younger generation did not tend to take over the role of voluntary observer as the founders of their departments retired: this role was then taken up by those who had a direct pecuniary interest in local weather conditions. As a result, by the 1870s, the more typical occupation of weather observer, according to Fleming, was that of farmer.

Conclusion

The contemporary weather observer in the Great Plains will see many similarities between weather conditions in the 1850s to the 1870s period and today. The fifteen years in which the Smithsonian Project functioned in Manhattan were sometimes dry, but commonly rather wet, and usually very warm in the summers. One difference lies in the record of astronomical phenomena, especially of the aurora borealis. Sunspots are the main reason given for their occurrence, but the utter darkness of the evenings at about nine o'clock, when the last of the daily observations were taken, is the other. Of course, there were no electric lights to dim the heavens. Another difference between the late 1800s and today is that the Kansas River rarely floods at Manhattan, at least not without the express permission of the Army Corps of Engineers.

The records of the Smithsonian Meteorological Project for Manhattan, Kansas have left us with both some basic weather data, and with a window into life on the American frontier as seen through the eyes of classically educated men with an interest in science. Their section of the Great Plains region was not a desert, nor could it be made into a lush, temperate garden. Although the daily average temperatures (32°F in January and 80°F in July: Self 1978) might seem to indicate a climate without a great deal of variation, they masked the wide daily fluctuations noted by the Smithsonian observers and the extreme weather that Mudge especially seemed to delight in recording. So these scientists were the first of their background to encounter the extremes of the Kansas climate, and recorded to their astonishment incidents of flooding, drought, a tornado, even an earthquake (Figure 3). Their records refute the claims of the immigrant guides, which had

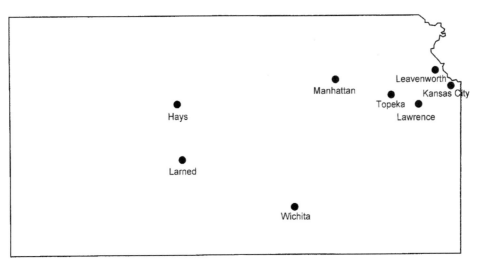

Figure 3. Map of Kansas.

contained overly optimistic claims of temperate weather conditions (Travis 1978). The Smithsonian Project records have indeed left us with a keen sense of nature's extremes on the Plains: this is their ultimate legacy for Manhattan, Kansas as well as for the surrounding region.

Acknowledgements

The author would like to thank Bruce Bradley and his staff of the Linda Hall Library, Mary Knapp, state climatologist for Kansas, and Nancy Hoyt, David Kromm, Doug Goodin, and John Harrington of the Department of Geography, Kansas State University, for their help on this paper. Thanks also to three anonymous reviewers for their useful comments, particularly about climate and climate change.

CHAPTER 10

CARBON SEQUESTRATION AND AGRICULTURAL GREENHOUSE GAS EMISSIONS IN CANADA BASED ON CEEMA ANALYSES

M.M. Boehm, S. Kulshreshtha, R.L. Desjardins and *B. Junkins*

ABSTRACT. Agriculture on the Canadian Prairies is a major economic activity and an important contributor of greenhouse gases (GHG). In 1990, prairie crop and livestock production contributed almost two-thirds of the greenhouse gas emissions from Canadian agriculture. GHG mitigation in agriculture is different than mitigation in other economic sectors of the region in two respects: besides carbon dioxide (CO_2) it emits large quantities of methane (CH_4) and nitrous oxide (N_2O) gases; and it can remove CO_2 from the atmosphere through Carbon (C) sequestration in agricultural soils. In this study, the magnitude of both the sources and removals (sinks) of GHG associated with agriculture were estimated to assess the relative potential of C sequestration to offset GHG emissions. An agroecosystem approach, embodied in the Canadian Economic and Emissions Model for Agriculture (CEEMA), was used to estimate emissions from 1990 agriculture, and to predict net emission levels in 2010 for business as usual and several mitigation strategies designed to enhance the soil sink potential. With business as usual by 2010, C sequestration in prairie agricultural soils could remove about 6 megatonnes (Mt) of CO_2, which is equivalent to about 13 percent of the emissions from prairie agriculture. However, adoption of zero tillage, forages in rotations, permanent cover cropping, improved management of grazing lands, and reduced summerfallow frequency could all further increase the amount of C sequestered. Zero tillage and improved management of grazing lands provided the largest opportunities for GHG reductions because their sink potentials were large enough to offset the increases in GHG emissions associated with their adoption. Both of the practices reduced direct GHG emissions from agriculture to below the 1990 level of emissions. If all of the mitigation measures were adopted by farmers, and agricultural soils were recognized as sinks, there is a good likelihood that agriculture's share of the commitments made by Canada under the Kyoto Protocol could be met.

Introduction

In 1997 the Canadian government made a commitment in the Kyoto Protocol to reduce national greenhouse gas (GHG) emissions by 6 percent below 1990 levels. If Canada ratifies the Protocol and it is internationally enforced, the GHG reduction must be achieved during the first commitment period from 2008 to 2012. To support the development of a national strategy for achieving the commitments made at Kyoto, the federal government established in 1998 the National Climate Change Secretariat (NCCS) and the Climate Change Action Fund (CCAF). Among the activities supported by the CCAF are Foundation Analysis

(Issues) Tables, which represent sectors of the Canadian economy or particular issues related to GHG and climate change. Sixteen Tables were given the task of producing a *Foundation Paper* and an *Options Report* in which policy options for GHG reductions in their sectors would be identified, analyzed, and evaluated. The Analysis and Modeling Group (AMG) which reports to the National Air Issues Coordinating Committee is responsible for the "roll-up" or cross-sectoral quantitative analysis of all of the Table options to assist in the development of the National Implementation Strategy.

Agriculture was represented in the national climate change process by the Agriculture and Agri-Food (AAF) Table. Its mandate was to provide CO_2, CH_4 and N_2O mitigation options for agriculture to the AMG. However, the AMG model is based mainly on CO_2 emissions from the use of fossil fuels, which is a major source of emissions in most sectors of the Canadian economy but only a small proportion of agricultural emissions. It was recognized by both the AAF Table and the AMG that a parallel modeling process that incorporated emissions of N_2O, CH_4 and CO_2 was required to understand and quantify mitigation strategies for agriculture. A two-step process was developed in which the AAF Table produced baseline and potential mitigation options for agriculture, and the Canadian Economic and Emissions Model for Agriculture (CEEMA) was used to quantify the resulting GHG emissions.

In this article, we summarize the results of the CEEMA analyses for options in which mitigation was based mainly on Carbon (C) sequestration in agriculture soils. Green-house gas reduction based on soil sinks is an important issue for the prairies, where the large agricultural land base and soil-conserving farming techniques used can be combined to offer a large potential for C sequestration.

The results of the CEEMA analyses for the full suite of GHG reduction strategies identified by the AAF Table are given in the Technical Report that accompanies the AAF Table Options Report (Junkins et al. 2000), who also provide more information about the sink-based scenarios, including cost curve analyses.

The major objectives of this study were: 1) to estimate agricultural GHG emissions for two baseline scenarios: 1990 and 2010 business as usual (BAU); and 2) to estimate GHG emissions for mitigation strategies based on enhancement of the C sequestration potential of agricultural soils.

CEEMA: The Analytical Framework

CEEMA represents the linkage of two sub-models: the Canadian Regional Agriculture Model (CRAM), an economic optimization model; and a Greenhouse Gas Emissions Model. It estimates agricultural emissions of the three major GHG: CO_2, N_2O and CH_4. Emissions levels for each of the gases are reported individually and as CO_2 equivalents (CO_2-Eq). The conversion to CO_2-Eq is based on 100-year warming potentials (1 for CO_2, 21 for CH_4, and 310 for N_2O); this makes it easier to compare emission levels among mitigation scenarios.

CRAM — The Economic Optimization Component

CRAM is an equilibrium model for Canadian agriculture that is disaggregated by commodity and region (Horner et al., 1992). When a change is introduced to the model, it solves for a new equilibrium position based on non-linear optimization that maximizes producer plus consumer surplus less transportation costs. The modeled commodities include grains, pulses and oilseeds, forages, beef, hogs, dairy, and poultry, which can be traded interprovincially and internationally in primary or processed form. Livestock and crop production are simulated for the 22 crop districts on the prairies and by province in the rest of Canada. Government policies are introduced through direct payments or indirectly through policies such as supply management and subsidized input costs.

Grain, oilseeds, and forage responses are determined by changes in the relative profitability of alternative crops. A calibration process duplicates the observed allocation of land by positioning an unobserved marginal cost curve such that conditions for constrained profit maximization are obtained. The marginal value product less the marginal cost for each output must equal the return to the fixed factor, i.e., land. At the margin, the return to land for each commodity is equal. The only constraint on crop production is the amount of land within each region (crop district or province), whereas beef and hog production reacts to changes in prices as well as input costs (such as the price of feed grains).

The model has been constructed so that demand cannot exceed available supply. As a result, CRAM can reach optimal solutions at less than full employment of resources if the returns are not expected to cover the variable costs of production.

Greenhouse Gas Emissions Component

CRAM determines optimal agricultural land use and levels of crop and livestock production for a given set of economic and market conditions. The GHG emissions component links the production levels to GHG emission coefficients and, based on a measure and multiply approach, calculates emissions. The emission coefficients are based on the current scientific knowledge and obtained from a variety of sources within Canada (Janzen et al. 1999; Environment Canada 1997; Statistics Canada 1997) and internationally (Houghton et al. 1997). The coefficients are described in detail in Junkins et al. (2000) and Kulshreshtha et al. (1999).

Eight potential sources of agricultural GHG emissions are represented in CEEMA and include:

> 1. Farm level crop production, which is represented by regional (crop district and provincial level) crop production systems. The cropping systems are defined by crop type, rotation, and tillage system. Potential sources of GHG emissions from crop production are crop residues, fertilizer application, fuel use, production of nitrogen-fixing crops, and changes in soil organic matter. An emission coefficient was assigned to each source, such that every crop-rotation-tillage system combination was represented by a unique level of emissions. Total emissions from the cropping systems were estimated by multiplying the level of activity of that system as generated by CRAM by the emission coefficients.

> 2. Emissions from farm level livestock production were direct emissions from farm animals, and manure handling, storage, and application. Emissions were estimated as for the crop production systems.

> 3. Emissions were estimated for the production of major farm inputs such as fertilizer, pesticides, petroleum products, and farm machinery, although under the IPCC accounting system they would be accounted in other sectors downstream from agriculture, such as manufacturing.

4. On-farm transportation represented transportation and storage of crop and livestock products, and activities related to the procurement of farm inputs. Although ideally they would be estimated separately, there were not sufficient data about on-farm fuel use to make distinctions among these activities. The emission coefficients and GHG emissions were estimated for each type of fuel used in on-farm transportation activities, including energy for drying grain crops and the augers that move grains into and out of storage.

5. Emissions from food processing were estimated for manufacturing processes and storage. The quantity and type of energy used for manufacturing were obtained from Hamilton, 1993. The food industry was disaggregated into meat and poultry, dairy products, fruits and vegetables, bakery products, vegetable oil mills, other foods, breweries, and other beverage industries.

6. Indirect nitrogen emissions were estimated for atmospheric deposition, leaching from fertilizer use, leaching from manure application in the field, cultivation of organic soils, and human sewage. The estimations followed the methodology of the IPCC (Houghton et al. 1997).

7. Other agro-ecosystem emissions included only farm shelter belts planted to reduce wind erosion, based on the IPCC guidelines (Houghton et al. 1997).

8. Off-farm transportation emissions were estimated by calculating the amount of product and the distance transported. For each product, energy use was allocated proportionally between trade and rail transportation modes.

According to the IPCC guidelines for the inventory of GHG, only direct emissions (and thus the first two sources on the preceding list) would be considered agricultural emissions. The rest are indirect emissions from agriculture, such as the production of N fertilizer or transportation of commodities from farm to market, and thus are accounted for in other sectors (i.e., manufacturing, transportation, or energy). The IPCC approach is suitable for GHG inventory, and was used by Neitzert et al. (1999) in the Canada's Greenhouse Gas Inventory. However, it cannot provide GHG emission information for the entire agricultural system, since it includes emissions which are interlinked with other sectors outside of the farm gate.

Carbon Sequestration Coefficients

The C sequestration coefficients developed for the CEEMA estimate the rate of C additions to the soil under specific tillage and cropping practices. However, there is large uncertainty about the rate of C sequestration in response to farming practices and land uses. The uncertainty occurs since it is difficult to measure changes in soil C resulting from differences in farming practices within relatively short periods (years rather than decades). The rate of gain or loss of soil organic C is small in relation to the total amount of organic C stored in soils, particularly prairie soils. For example, the organic C content of soils ranges from about 100 tonnes per hectare in the Black soil zone to 60 tonnes per hectare in the Brown soil zone (Anderson 1995), whereas annual additions resulting from the adoption of soil-conserving farming practices are in the range of 0.1 to 0.5 tonnes C per hectare

Table 1. Carbon Sequestration Using "Expert Opinion" and "Century" Coefficients (t $CO_2ha^{-1}yr^{-1}$)

Activity	Prairies Soil Zone						Non-Prairie	
	Brown		Dark Brown		Black			
	Expert Opinion[1]	Century[2]	Expert Opinion	Century	Expert Opinion	Century	Expert Opinion	Century
Zero tillage[3]	0.73	0.22	0.73	0.44	1.34	0.54	0.76	0.54
Minimum tillage		0.08		0.16		0.26		0.26
Reduce SF [4]	0.15	0.13	0.16	0.29	0.08	0.20		
Convert to forage or pasture	0.73		1.78	0.94	3.23	2.44	3.23	2.44

1. McConkey et al. 1999; 2. Smith et al. 2000; 3. The Expert Opinion value for zero tillage in non-prairie regions was taken from Century output, June 1999; 4. The Expert Opinion SF coefficients are calculated from cropping frequency and vary among scenarios. Shown here are the coefficients for the 2010 baseline.

(Bruce et al. 1999). In addition, the spatial variability of soil organic C within a field or region is often greater than the annual rate of change in response to management practices, further complicating the detection of systematic trends in soil organic C storage.

Despite the difficulties in precisely measuring and predicting changes in soil C content, the conditions under which it is possible to sequester C are understood (i.e., Janzen et al. 1999). The adoption of farming practices that reduce soil disturbance while increasing biomass and organic C will result in an increase in soil C.

Two sets of C sequestration coefficients were developed for the CEEMA (the "Expert Opinion" and "Century") and are presented in Table 1. The Century coefficients were developed for CEEMA by Smith et al. (2000) using their Century model that simulates the rate of C change associated with changes in crop-rotation-tillage combinations. When the coefficients were reviewed by the AAF Table, some members suggested that there were sufficient data on C storage for adoption of zero tillage and reduced summer fallow frequency on the Prairies that empirical coefficients could be developed. McConkey et al. (1999) developed the "Expert Opinion" coefficients from both empirical data and expert opinion. All of the sink-enhancing mitigation scenarios were analyzed using both set of coefficients, resulting in two estimates of the sink potential per scenario.

The C sink was quantified for each land use or tillage system by multiplying the area of land under that system with its associated coefficient from Table 1. The exception is the calculation of the rate of C sequestration in response to a decline in summerfallow frequency using the Expert Opinion coefficient. In that case, the amount of C sequestration was based on the change in cropping frequency over the time of the analysis, such that the greater the reduction in summerfallow acreage per unit time, the greater the C sequestration potential (McConkey et al. 1999) so that:

$$C_{gain} = C_{no\ fallow}\ (CF_{2010} - CF_{1990}) \tag{1}$$

where: $C_{no\ fallow}$ (Mg C ha^{-1} yr^{-1}) = 0.4 in the Brown soil zone and 0.6 in the Dark Brown, Black, and Gray soil zones

CF_{1990} = cropping frequency per crop district in 1990

CF_{2010} = cropping frequency per crop district in 2010.

For each region (crop district or province), the change in cropping frequency and C sequestration coefficient were calculated, and the coefficient was multiplied by the hectares of cropland in that region.

Aggregation of the GHG Emissions

Greenhouse gas emissions were estimated for three levels of the agricultural system:

> 1. IPCC agriculture emissions, as described in the Guidelines for GHG Inventory (Houghton et al. 1997). These represent direct emissions from primary crop and livestock production activities.

> 2. Total primary agriculture emissions, which are IPCC agriculture plus on-farm energy use and soil sinks. On-farm energy includes fossil fuel use for crop production activities, such as cultivation, seeding, and crop harvest.

> 3. Total agriculture, which is total primary agriculture emissions plus transportation, food processing, agro-ecosystem and other indirect emissions.

Results of the scenario analyses are reported for each level of aggregation. The IPCC agriculture emissions show emissions within the context of the Kyoto Protocol, whereas total primary agriculture and total agriculture emissions demonstrate how the changes proposed in the scenarios will affect GHG emissions within the broader Canadian economy.

Greenhouse Gas Emissions from Agriculture

Estimates of agricultural GHG emissions and the C sink were made for eight scenarios: 1990 baseline; 2010 BAU; increased zero tillage; reduced summerfallow frequency; increased forages in crop rotations; two permanent cover scenarios; and improved grazing management. The 1990 baseline and 2010 BAU scenarios provide the benchmark against which the sink-enhancing mitigation scenarios are measured.

The Baseline Scenarios

1990 Baseline

The CRAM component of CEEMA was calibrated with 1991 Census of Agriculture data for crop production activities in 1990. Total agricultural land was about 45 million ha, of which 35 million ha (78 percent) was in cropland with the remaining 10 million ha in hayland and pasture. The prairies accounted for about 85 percent (38.3 million ha) of Canadian agricultural land, including 90 percent of the cropland and about 7.8 million ha of summerfallow. The regional distribution of tillage practices (conventional, minimum, and zero tillage) was estimated using the 1991 Census of Agriculture. Further, the 1991 Census estimates that in Canada there were about 4 million beef cows, 1.3 million dairy cows, 15 million hogs and 400 million poultry broilers. A majority of the beef cattle (77 percent), but only 38 percent of the hogs, 15 percent of the dairy cows, and only a small proportion of the poultry were produced on the Canadian Prairies.

In 1990, Canadian GHG emissions from IPCC agriculture were 57.6 Mt of CO_2-Eq (Table 2). Livestock production was the largest source of emissions, mainly due to CH_4 and N_2O. Sixty-four percent of the direct emissions occurred in the prairies

Table 2. Greenhouse gas emission scenarios for agriculture in Canada and the Canadian Prairies from baseline in 1990 to Business as Usual 2000

Source	Prairies				Canada				change as % of 1990
	CO_2	CH_4	N_2O	CO_2-Eq	CO_2	CH_4	N_2O	CO_2-Eq	
1990 baseline									
I.P.C.C. Agriculture	5,693	517	65	36,544	6,033	994	99	57,574	0.0
Sink Offset	0			36,700	-10			57,564	0.0
Total Primary Agriculture	11,356	517	66	42,550	13,680	994	100	65,670	0.0
Total Agriculture & Food	22,405	751	73	60,788	30,118	1,265	109	90,466	0.0
2010 business-as-usual – Century coefficients									
I.P.C.C. Agriculture	265	692	94	43,837	563	1,138	131	64,962	11.3
Sink Offset	-5,515			38,422	-5,826			59,136	2.5
Total Primary Agriculture	330	693	95	44,258	2,407	1,139	132	67,270	2.3
Total Agriculture & Food	14,490	947	103	66,291	23,947	1,432	142	97,971	7.2
2010 business-as-usual – Expert Opinion coefficients									
I.P.C.C. Agriculture	148	692	94	43,820	537	1,138	131	64,936	11.3
Sink Offset	-5,896			37,924	-6,280			58,656	1.9
Total Primary Agriculture	-169	693	95	43,750	1,927	1,139	132	66,790	1.0
Total Agriculture & Food	13,992	947	103	65,790	23,467	1,432	142	97,491	6.7

I.P.C.C. (Intergovernmental Panel on Climate Change) Agriculture is direct emissions from crop and livestock production activities; the Sink Offset is I.P.C.C. Agriculture minus the soil C sink; Total Primary Agriculture includes sinks and on-farm fuel use; and Total Agriculture and Food includes indirect and food-processing emissions. Percentage changes are relative to the 1990 baseline.

(Table 2). Agricultural soils in Canada were a net source of CO_2 with emissions of 6 MT compared to a C sink of only 10 kT (Table 2). Prince Edward Island was the only province in which soils were a net sink of CO_2. The total primary agricultural emissions for Canada were 65.7 Mt CO_2-Eq. Emissions from the total agriculture and food system were 90.5 Mt of CO_2-Eq (Table 2). At all levels of aggregation, estimated emissions occurred in proportion to the area of cropped land and the size of the animal herd in each province. The prairie provinces, with the largest land base and animal herd in Canada, produced the most agricultural emissions.

2010 Business as Usual

The Medium Term Baseline (MTB) projections given in Agriculture and Agri-Food Canada (1999a) provided the basis for predictions of BAU agriculture activities to 2010. Land management practices, such as summerfallow and zero tillage frequency, were based on historical trends established from 1981 to 1996 using Census of Agriculture data.

More information on the 2010 BAU scenario is available in Junkins et al. (2000), but the major assumptions are here summarized as follows: agricultural cropland for Canada was held constant at the 1996 level of 34.2 million ha in crops and 10.8 million ha in hay, forages and pasture; crop and hay yields were increased on the linear historical trend-line; on the prairies N fertilizer use increased by 25 percent over the 1996 level, but was held at 1996 levels in Eastern Canada; total area in summerfallow was reduced to 5 million ha; zero tillage averaged about 30 percent (8.8 million ha) on the prairies, although there was considerable regional variability; and animal numbers were increased based on the MTB trends, with a larger proportion of the increase occurring in western Canada. Beef numbers increased by 10 percent in the west and 2 percent in the east, and hogs increased by 31 percent in the west and 8 percent in the east.

Although cropland was held at 1996 levels, crop production increased between 1990 and 2010 owing to reduced summerfallow frequency and the projected increase in yields. Beef production increased by 34 percent and hog production increased by 43 percent of 1990 levels.

BAU IPCC agriculture emissions were 65 Mt CO_2-Eq in 2010, an increase of 11.3 percent above the 1990 baseline emissions. Prairie agriculture produced 44 Mt CO_2-Eq of emissions, i.e. 65 percent of national emissions. The increase occurred mainly as CH_4 and N_2O emissions from the expanded livestock herd and increased N fertilizer use on the prairies (Table 2). Prairie agriculture accounted for 52 percent of the CH_4 and 65 percent of the N_2O emissions in 1990, compared to 60 percent of the CH_4 and 72 percent of the N_2O in 2010. The prairies would have accounted for a larger proportion of agricultural emissions, but CO_2 emissions from soil organic matter decomposition declined from 6 to 0.5 Mt between 1990 and 2010 as zero tillage practices increased and summerfallow frequency decreased (Table 2). At the same time, the amount of C sequestered in prairie soils rose to ~5.5 Mt CO_2 (Table 2). The Canadian soil sink was 5.8 Mt CO_2, resulting in total primary agricultural emissions of 67 Mt CO_2-Eq, which is 2.4 percent higher than in 1990 (Table 2).

Sink Enhancing Scenarios

The sink enhancing scenarios were based on the assumption that soils reach organic C equilibrium after decades of use in a particular production system. A shift in production practices will cause a gradual change in the soil organic carbon (SOC) content until equilibrium is achieved for the new system, after which there will be no net gain or loss. Soils will gain C if biomass inputs (i.e., crop residue, manure) exceed the rate of soil C loss (i.e., decomposition). Management factors that directly influence C inputs are fallow frequency and nutrient inputs either from legume and pulse crops or fertilizers. In production systems with large inputs of biomass and low rates of organic matter mineralization, soil C levels will increase toward a new equilibrium level for that system.

The success of the scenarios cannot be determined solely on the basis of the size of the C sink. Some scenarios cause an increment increase, not only in the sink, but also in GHG emissions. They must therefore be assessed on the basis of "net" emissions, in which the total sink is subtracted from total emissions. In this report, scenario outcomes are assessed in terms of 'net' emissions rather than in terms of the "gross" C sink.

Increased Utilization of Zero Tillage Practices

In all soil conserving farming systems, there are two common factors that prevent organic matter loss. The first is the addition to the soil of sufficient plant residues and nutrients to maintain or enhance the soil organic C level; the second is sufficient protection of the soil by crop, stubble or permanent plant cover to prevent erosion by wind and water (Wood and Edwards 1992). This is accomplished by reducing summer fallow and tillage intensity (Campbell et al. 1991; Wood and Edwards 1992), while adding adequate nutrients (N, P, and K) to the soil (Janzen 1987a).

Many researchers have measured increases in the organic matter content of the Ap soil horizon where no-till practices were used (e.g. Bruce et al. 1990; Havlin et al. 1990; Carter 1992; Wood and Edwards 1992; Campbell and Zentner 1993; Mahboubi et al. 1993; Beare et al. 1994b). These researchers attribute the increased organic matter to reduced erosion, to higher yields resulting in more crop residue added to the soil surface, and to differences in the assimilation and decomposition of soil organic matter.

In zero tillage systems, the soil gradually becomes physically and chemically stratified, with a mulch of accumulated plant litter, rich in organic C and nutrients at the soil surface (Wood and Edwards 1992; Beare et al. 1994; Gregorich et al. 1994; Lal et al. 1994). The mulch layer tends to insulate the soil from temperature extremes and rapid desiccation and creates a stable environment for microbial activity (Campbell 1991; Carefoot et al 1990; Hendrix et al 1988). The microbial ecology of these soils are more like those of natural grasslands (Beare et al. 1994a). Thus there is a lower rate of biological oxidation, so a large proportion of the organic mulch is eventually converted to stable soil organic matter (Gallaher and Ferrer 1986).

To estimate the GHG mitigation value of greater use of zero tillage systems, an expanded zero tillage scenario was used. Its adoption was assumed to be 50 percent (14.7 million ha) by 2010 on cropland in the prairies and Peace River region of British Columbia (BC), compared to 30 percent for the 2010 BAU scenario. Nitrogen fertilizer use was increased by 5 percent over the BAU scenario to ensure that crop biomass production and C sequestration were not constrained by a nitrogen limitation (Halvorson et al. 1999). The mix of crops produced on the prairies did not change as a result of the increase zero tillage, and there were no changes in livestock production. The rate of C sequestration associated with zero tillage practices was estimated using the Expert Opinion and Century coefficients given in Table 1.

GHG emissions were 12 percent greater under the expanded zero tillage scenario than in 1990 (Table 3), mainly owing to higher emissions associated with increased N fertilizer use and cattle production. For both Canada and the prairies, the C-sink calculated with the Century coefficients was ~40 percent larger (8.1 Mt CO_2) than in 2010 BAU (5.8 Mt CO_2) (Table 3). The largest sink (5.6 Mt CO_2) occurred in Saskatchewan, where there was more potential for the adoption rate of zero tillage. "Net" emissions for Canada (IPCC agriculture minus the sink offset) were 0.5 percent lower than the 1990 baseline (Table 3). Total primary agriculture emissions, which include both the sink and on-farm fossil fuel use, were 0.7 percent lower than in 1990. Total agriculture and food system emissions were 96.5 Mt CO_2, an increase of 6.2 percent relative to 1990 (Table 3).

The Expert Opinion coefficients for C sequestration under zero tillage were larger than the Century coefficients. As a result, for Canada the soil sink was 11.8 Mt CO_2-Eq, compared to 8.1 Mt CO_2-Eq with the Century coefficients (Table 3). Net emissions (IPCC agriculture minus the sink offset) were 53.5 Mt CO_2-Eq, a reduction of 7.6 percent below the 1990 baseline. Because of the large C sink, emissions from total primary agriculture were 7 percent below the 1990 baseline; and total agriculture and food emissions were only 2.4 percent greater than in 1990.

The shift of land from conventional tillage to zero tillage altered the mix of crop production inputs. Weed control was achieved with herbicides rather than tillage; and more N fertilizer was used, which increased the use of chemical inputs while reducing the amount of machinery and fossil fuel use. The decline in total primary agriculture emissions relative to 1990 reflects the large C sink coupled with

Table 3. Canada and Canadian Prairies GHG agricultural emissions in 2010 for increased zero tillage and reduced summerfallow scenarios.

Source	Emissions (kt)								change as % of 1990
	Prairies				Canada				
	CO$_2$	CH$_4$	N$_2$O	CO$_2$-Eq	CO$_2$	CH$_4$	N$_2$O	CO$_2$-Eq	
Increased zero tillage – Century coefficients									
I.P.C.C. Agriculture	149	692	95	44,239	447	1,138	132	65,360	11.9
Sink Offset	-7,776			36,355	-8,086			57,273	-0.5
Total Primary Agriculture	-2,234	693	96	42,201	-158	1,139	134	65,210	-0.7
Total Agriculture & Food	12,265	943	104	64,445	22,075	1,428	143	96,480	6.2
Increased zero tillage – Expert Opinion coefficients									
I.P.C.C. Agriculture	0	692	95	44,089	389	1,138	132	65,302	11.8
Sink Offset	-11,422			32,560	-11,806			53,496	-7.6
Total Primary Agriculture	-6,030	693	96	38,406	-3,955	1,139	134	61,432	-6.9
Total Agriculture & Food	8,470	943	104	60,649	18,298	1,428	143	92,702	2.4
Reduced summerfallow – Century coefficients									
I.P.C.C. Agriculture	241	707	99	45,712	534	1,147	135	66,597	13.5
Sink Offset	-5,689			40,135	-6,120			60,655	5.0
Total Primary Agriculture	18	707	100	45,853	2,114	1,148	137	68,648	4.3
Total Agriculture & Food	14,654	964	108	68,440	24,314	1,443	147	100,100	9.6
Reduced summerfallow – Expert Opinion coefficients									
I.P.C.C. Agriculture	18	709	99	45,540	399	1,148	135	66,480	13.4
Sink Offset	-8,854			36,722	-9,185			57,295	-0.5
Total Primary Agriculture	-3,205	709	100	42,681	-1,072	1,149	137	65,480	-0.3
Total Agriculture & Food	11,409	966	108	65,257	21,065	1,444	147	96,880	6.3

I.P.C.C. (Intergovernmental Panel on Climate Change) Agriculture is direct emissions from crop and livestock production activities; the Sink Offset is I.P.C.C. Agriculture minus the soil C sink; Total Primary Agriculture includes sinks and on-farm fuel use; and Total Agriculture and Food includes indirect and food-processing emissions. Percentage changes are relative to the 1990 baseline.

a decrease in fossil fuel use, whereas the increase in total agriculture and food emissions reflects a rise in the manufacture and transport of crop production inputs and commodities.

Although there is uncertainty associated with both the C sequestration coefficients and adoption rates, the results suggest that the adoption of zero tillage practices on the prairies offers significant potential for C sequestration. However, if C removals are not recognized in the Kyoto Protocol and the C offset cannot be used, adoption of zero tillage will increase IPCC agriculture emissions when compared to the baseline scenario.

Decreased Utilization of Summerfallow

In the arid regions of the prairies, summerfallow is used to store soil moisture, control weed growth, and increase the rate of nutrient release stored in soil organic matter. Because much of the Canadian Prairies has a growing-season moisture deficit, summerfallow reduces the risk of crop failure due to drought while minimizing the costs of chemical inputs. However, summerfallow is directly related to losses in soil organic matter. In fallow years, the risk of soil erosion is high because it is not protected by plant cover. Further, organic C losses result because the rate of decomposition is accelerated in the moist, warm surface soil and plant residue additions to the soil are low (Doran and Linn 1994).

Many prairie farmers have reduced summerfallow frequency since this reduces soil degradation, improves profitability, and conserves energy. Increased cropping frequency reduces organic matter loss and provides good economic returns, although it makes yield more variable (Larney et al. 1994).

In the reduced summerfallow scenario, summerfallow frequency in the prairie region was reduced by 50 percent (from 1.4 to 0.7 million ha) in the Black soil zone, by 40 percent (1.4 to 0.8 million ha) in the Dark Brown soil zone, and by 30 percent (2.7 to 1.5 million ha) in the Brown soil zone. Average crop yields declined, reflecting the generally lower yields of crops seeded on stubble compared to fallow. However, since the total area of land seeded to crops increased as summerfallow acreage declined, total crop production was higher. The mix of crops remained about the same as for the 2010 BAU scenario, except that hay production on the prairies increased as some summerfallow land was converted to forage. As a result of the increase in feed, livestock production increased by about 2 percent. Direct GHG emissions from crop production and indirect emissions from the manufacturing and transportation of crop inputs and commodities increased in proportion to the increase in cropped land.

The Century model coefficients for C sequestration resulting from reduced summerfallow frequency are shown in Table 1. The Expert Opinion coefficients were calculated from the change in cropping frequency predicted by CRAM as a result of the scenario constraints. The coefficients in t CO_2 ha^{-1}yr^{-1} were 0.33 in the Brown soil zone, 0.23 in the Dark Brown soil zone, and 0.15 in the Black and Gray soil zones. These coefficients were multiplied by the hectares of cropland for each soil zone represented within a crop district in order to estimate the sink potential from reduced summerfallow.

Canadian IPCC agriculture emissions rose by ~13% above the 1990 baseline, and about 2 percent above 2010 BAU (Table 2). The increase was mainly N_2O emissions from N fertilizer, which rose in proportion to the increase in seeded area (Table 3). The soil C sink predicted with the Century coefficients was 6.0 Mt CO_2; and net emissions (IPCC agriculture emissions minus the sink offset) were 60.7 Mt CO_2-Eq, an increase of 5 percent from the 1990 baseline (Table 3) and 2.5 percent from 2010 BAU. Total primary agriculture emissions were 4.3 percent greater, and total agriculture emissions were 9.6 percent greater than in 1990 (Table 3).

The Expert Opinion coefficients predicted a sink of 9.2 Mt in Canada, much of which (8.9 Mt CO_2) was for the prairies. The sink offset reduced IPCC agriculture emissions to 57.3 Mt CO_2-Eq, which is 0.5 percent lower than in 1990. Total primary emissions were about the same, and total agriculture emissions were about 6 percent greater than in the 1990 baseline scenario (Table 3).

The mitigation value of reduced summerfallow frequency based on these analyses is uncertain, as illustrated by the difference in outcome depending on the C sequestration coefficient values. Emissions are the same as 2010 BAU using reduced summerfallow frequencies (indicated by the Century coefficients) even with the C sink offset. If the sink potential predicted by the Expert Opinion coefficients can be achieved, this strategy would offer a slight reduction in emissions relative to the baseline scenarios.

Increased Use of Forage in Rotations with Grains and Oilseeds

The inclusion of legume forages in grain and oilseed rotations has the potential to reduce GHG emissions through enhanced C sequestration and reduced use of N fertilizer. Carbon storage is enhanced because forages increase the amount of biomass added to the soil and N fertilizer use declines because legume forages add biologically fixed N to the soil. Increased legume forage production will also increase emissions, however. Legumes are themselves a direct source of N_2O, and an increase in the supply of forages may trigger an expansion of the livestock herd, which in

Table 4. Changes in productivity and N fertilizer associated with the increase in forages in rotation.

Parameter	Crop	Hayland:cropland (%)		
		<10	10-25	>25
Productivity	Grains and oilseeds (% change)	10	5	2
	Hay (% change)	10	5	2
N fertilizer	Grains and oilseeds (% change)	-20	-8	-4

Table 5. Canada and Canadian Prairies GHG agriculture emissions in 2010 for increased forage and permanent cover scenarios.

Source	Emissions (kt)								change as % of 1990
	Prairies				Canada				
	CO_2	CH_4	N_2O	CO_2-Eq	CO_2	CH_4	N_2O	CO_2-Eq	
Increased forages in rotation – Century coefficients									
I.P.C.C. Agriculture	159	1,057	109	56,151	492	1,486	145	76,543	24.8
Sink Offset	-8,453			47,698	-8,731			67,811	15.1
Total Primary Agriculture	-2,242	1,058	110	54,145	-113	1,487	146	76,439	14.1
Total Agriculture & Food	11,674	1,345	119	76,925	21,704	1,812	157	108,502	16.7
Permanent cover program, no cattle increase – Century coefficients									
I.P.C.C. Agriculture	19	703	93	43,680	319	1,140	129	64,372	10.6
Sink Offset	-6,064			37,547	-6,308			58,064	0.8
Total Primary Agriculture	-548	704	94	42,465	1,596	1,141	131	66,108	0.7
Total Agriculture & Food	13,494	957	102	65,347	23,070	1,432	141	96,718	6.5
Permanent cover program, no cattle increase – Expert Opinion coefficients									
I.P.C.C. Agriculture	12	703	93	43,687	405	1,140	129	64,543	10.8
Sink Offset	-7,378			36,227	-7,663			56,880	-1.2
Total Primary Agriculture	-1,869	724	94	42,159	327	1,141	131	64,924	-1.1
Total Agriculture & Food	12,173	957	104	64,040	21,801	1,432	141	95,534	5.3
Permanent cover program, increase in cattle – Century coefficients									
I.P.C.C. Agriculture	164	724	94	44,576	468	1,168	131	65,587	12.2
Sink Offset	-6,211			38,297	-6,521			59,066	2.5
Total Primary Agriculture	-527	724	95	44,239	1,554	1,169	132	67,135	2.2
Total Agriculture & Food	13,532	978	104	66,163	23,005	1,462	142	97,744	7.4
Permanent cover program, increase in cattle – Expert Opinion coefficients									
I.P.C.C. Agriculture	11	724	94	44,437	409	1,168	131	65,612	12.3
Sink Offset	-7,387			36,968	-7,772			57,841	-0.5
Total Primary Agriculture	-1,856	724	95	42,923	245	1,169	133	65,910	0.4
Total Agriculture & Food	12,273	978	104	64,847	21,696	1,462	142	96,519	7.4

I.P.C.C. (Intergovernmental Panel on Climate Change) Agriculture is direct emissions from crop and livestock production activities; the Sink Offset is I.P.C.C. Agriculture minus the soil C sink; Total Primary Agriculture includes sinks and on-farm fuel use; and Total Agriculture and Food includes indirect and food-processing emissions. Percentage changes are relative to the 1990 baseline.

turn will increase emissions of CH_4 and N_2O. The mitigative value of the scenario thus depends on the net balance between emissions reductions and increases.

The forage rotation scenario was based on an increase in forage production in crop rotations of 2.8 million ha in the Prairie and Peace River regions. The increase was allocated at the crop district level in proportion to the amount of forage production reported in the 1996 census. Yields of grains and oilseeds were assumed to rise, and the N requirement from fertilizer was assumed to decline in response to biologically-fixed N added to soils by the legume forages. Productivity was increased by 10 percent and N fertilizer use decreased by 20 percent in crop districts with only a small proportion of hayland (relative to cropland), where it was assumed that the inclusion of forages would significantly improve soil quality (Table 4). Conversely, where the hayland:cropland ratio was high (>25 percent), it was assumed that grain and oilseed yields already reflect the benefits of forages in rotation, so the yield increase was only 2 percent and N fertilizer use was reduced by 4 percent (Table 4). Livestock numbers were increased to utilize the supply of forage that resulted from the scenario.

Carbon sequestration was 30 percent greater in this scenario than the 2010 BAU scenario, but the increase in direct emissions from livestock and legumes were incrementally larger. Direct emissions from livestock increased by 54 percent, emissions from nitrogen-fixing crops increased by 27 percent, and overall IPCC agriculture emissions increased by 18 percent. Compared to 1990, emissions from IPCC agriculture were 25 percent higher, although total primary agricultural emissions were only 14 percent higher because of the large C sink and lower on-farm fossil fuel use for legume forage production compared to grains and oilseed production (Table 5).

This scenario was not tested with the Expert Opinion Coefficients because of its low mitigation potential.

Expansion of the Permanent Cover Program on the Prairies

The major mitigation goal of an expanded permanent cover program (PCP) is to sequester C; it was assumed that perennial cropping would lower the rate of organic matter decomposition and increase the rate of biomass additions to enhance the soil C sequestration potential. The PCP scenario was developed by the Prairie Farm Rehabilitation Administration (PFRA) to convert 1 million ha of marginal cropland to permanent cover on the basis of the current extent of cultivated marginal land, hayland, and pasture within each crop district. The rate of conversion was greater in crop districts that had the highest proportion of marginal cropland and enough hayland and pasture to support cattle production: livestock production thus increased where it was already a major agricultural activity.

The conversion to permanent cover involved 167,000 ha in Manitoba, 461,000 ha in Alberta and 417,000 ha in Saskatchewan. Most of the permanent cover land (880,000 ha) shifted into tame pasture, with the balance in hay production. The rate of C sequestration associated with the conversion of cropland to forage and hay production was estimated using the coefficients in Table 1.

Two expanded PCP scenarios were evaluated. Scenario PCP1 assumed that the area of permanent cover crops increased by 1 million ha but that there was no corresponding increase in the size of the cattle herd. The second scenario (PCP2) allowed cattle numbers to increase in order to utilize unimproved pastureland to the same extent as in the 2010 BAU scenario.

PCP1: No increase in cattle

The conversion of 1 million ha of marginal cropland to improved pasture and hayland without a corresponding increase of cattle numbers to utilize the increased feed supply resulted in a 24 percent increase in pasture land and 1.87 million ha of ungrazed pasture. Annual crop acreage declined by 6.2 percent nationally, and by 20 percent in Alberta, 2 percent in Saskatchewan, and 4 percent in Manitoba.

The expansion of permanent cover resulted in an overall 11 percent increase in IPCC emissions from Canadian agriculture (Table 5). Net IPCC emissions (IPCC agriculture minus the sink) and total primary agriculture emissions ranged from a ~1 percent increase (Century Coefficients) to a ~1 percent decrease (Expert

Opinion Coefficients), compared to the 1990 baseline (Table 5). Although total agriculture emissions increased relative to 1990, they were lower than in the 2010 BAU scenario, mainly because crop production inputs declined and sequestered C increased as cropland shifted to pasture and hay production.

PCP2: Increase cattle numbers to utilize available pasture

Permanent cover acreage was expanded and allocated as in the previous scenario, but the cattle herd increased by ~6 percent in order to graze the available pasture. Direct crop emissions decreased by 2 percent, mainly owing to decreases in CO_2 and N_2O emissions from soils and fertilizer use respectively. Direct livestock emissions increased by 3 percent, mainly as CH_4. The C sink increased by about 1 Mt CO_2, so overall net emissions were similar to business as usual in 2010. Compared to 1990, IPCC agriculture emissions minus the sink ranged from 2.5 percent higher with the Century Coefficients to 0.5 percent lower with the Expert Opinion Coefficients (Table 5). Whether this scenario is mitigative depends on the magnitude of the C sink in the pasture soils.

Improved Grazing Management Scenario

Three combined grazing management strategies were examined to look at cattle grazing and carbon emissions. Using information provided by Martin et al. (1999), three grazing management strategies were combined in this scenario: 1) reduction in the stocking rate of native rangelands; 2) complementary grazing; and 3) rotational grazing. Adoption rates of these strategies are shown in Table 6.

The grazing scenario estimated the GHG emissions associated with improved management of native ranges, and native and tame pasturelands. The improved management of grazing lands both enhanced the soil sink potential and the efficiency of the livestock production system. The quality and yield of forages and pastures were assumed to increase, thus improving calving rates, weaning weights and overall cattle productivity, while reducing GHG emissions per animal.

To achieve the reduced stocking rate on native prairie pastures, the model took a proportion of the cattle herd off native rangelands and fed them on feedlots. Hay acreage was increased to meet the demand for livestock feed. The quality of forage was assumed to increase, thereby lowering net grain and forage feed requirements.

To prevent overgrazing, complementary grazing was applied to native rangelands in the moist regions of the prairies. Crested wheatgrass was grazed from May 16 to July 9; native range from July 10 to August 19; and Russian wild rye from August 20 to October 31. Complementary grazing optimizes forage use when the pasture and rangelands are in prime quality. As a result, calving rates and cattle weaning weights were higher, and grain and forage feed demands declined. The cost of seeding crested wheatgrass and wild rye on native rangeland was represented by shifting about 75 percent of the required rangeland area into improved pasture.

The rotational grazing strategy involved moving cattle between subdivided sections of pasture in order to harvest the forages at their peak; it was applied to tame pastures in Western Canada and to natural pastures in Eastern Canada. A C sequestration benefit results because the risk of overgrazing is reduced and manure additions are more effectively utilized. Rotational grazing with legumes improves soil nitrogen levels, thereby increasing productivity and the C sequestration potential. However, the costs of fencing, posts, watering and seeding increased production costs of this strategy as compared to the 2010 BAU baseline.

Changes in the crop mix and yields of grains and oilseeds were minor because of counteracting signals resulting from the combined strategies. Whereas improved forage quality lowered the demand for feed grains, lower stocking rates on native rangeland increased the demand for feed grains, so that net change was inconsequential. The extent of tame pasture increased by 13 percent, and unimproved pasture declined by 11 percent, mainly in the prairies.

Total IPCC agricultural emissions for the combined grazing strategies were about the same as for the 2010 BAU baseline (Table 7). Even though cattle numbers rose by 6 percent, which increased CH_4 and N_2O emissions compared to BAU, the amount of C sequestered under the improved grazing management was sufficient to offset the increase in emissions from cattle. IPCC emissions minus the sink and total primary agriculture emissions were ~2 percent (Century Coefficients) to ~3 percent (Expert Opinion Coefficients) lower than in the 1990 baseline (Table 7). Total agriculture emissions declined relative to 2010 BAU, but were ~5 percent

Table 6. Adoption rates used in the improved grazing management scenarios.

		Adoption Rates (%)	
Parameter	Regions and Assumptions	2010 BAU	Grazing Scenario
Reduced stocking rate	British Columbia, western Manitoba, northern Saskatchewan, Alberta	25	35
Complimentary grazing	British Columbia, western Manitoba, northern Saskatchewan, Alberta	25	35
Rotational grazing	Western and central Canada tame pasture; legumes in rotation, no N fertilizer, re-seed every 5 years	0	10
	Eastern tame pasture: maintain legumes in rotation, not N fertilizer, manure provides nutrients	0	5
	Eastern natural pasture: add P, K and lime, no legumes	0	10

Table 7. Canada and Canadian Prairies GHG agriculture emissions for the grazing management scenarios.

	Emissions (kt)								
Source	Prairies				Canada				change as % of 1990
	CO_2	CH_4	N_2O	CO_2-Eq	CO_2	CH_4	N_2O	CO_2-Eq	
Grazing Management strategies - Century coefficients									
I.P.C.C. Agriculture	253	701	94	44,160	253	1,149	131	64,969	11.4
Sink Offset	-7,141			36,973	-8,471			56,498	-1.9
Total Primary Agriculture	-1,442	702	95	42,823	-532	1,149	132	64,650	-1.6
Total Agriculture & Food	12,751	957	103	64,917	21,071	1,444	142	95,454	5.2
Grazing Management strategies - Expert Opinion coefficients									
I.P.C.C. Agriculture	170	701	94	44,077	70	1,149	131	64,785	11.1
Sink Offset	-7,094			36,942	-8,888			55,897	-3.0
Total Primary Agriculture	-2,165	702	95	42,099	-1,132	1,149	132	64,049	-2.5
Total Agriculture & Food	12,027	957	103	64,193	20,471	1,444	142	94,853	4.6

I.P.C.C. (Intergovernmental Panel on Climate Change) Agriculture is direct emissions from crop and livestock production activities; the Sink Offset is I.P.C.C. Agriculture minus the soil C sink; Total Primary Agriculture includes sinks and on-farm fuel use; and Total Agriculture and Food includes indirect and food-processing emissions. Percentage changes are relative to the 1990 baseline.

greater than 1990 baseline emissions. The major factor responsible for the decline in total agriculture emissions relative to the 1990 baseline was the large C sink in the soil. This sink was estimated to be 30 percent greater than the 2010 BAU scenario as a result of improved management of grazing lands.

Discussion

Analyses of GHG mitigation based on enhanced C sequestration has demonstrated the potential of agricultural soil sinks to offset GHG emissions. In its Options Report, the Agriculture and Agri-Food Table singled out technologies that sequester C as the best options for "verifiable low cost reduction in GHG emissions" in agriculture (AAF Table, 2000, p. 27). The Agriculture and Agri-Food Table found, on the basis of a range of GHG mitigation strategies, that if soil sinks cannot be included in the emissions inventory, it will be very difficult for prairie (or Canadian) agriculture to achieve GHG emissions reductions in the order of 6 percent below 1990 levels. Other technologies, such as improved nutrient management, livestock feeding or manure management, may offer long-term potential for GHG reductions, but they presently are not yet well enough understood or developed to be promoted actively (AAF Table, 2000). In contrast, many of the C sequestering practices are already widely used (i.e., zero tillage, reduced summerfallow frequency, or conversion of marginal cropland to permanent cover) and are understood well enough to be promoted. Other land uses and management practices, such as the restoration of wetlands or agro-forestry, may even offer larger C sink capacity than the strategies reported in this article. However, an understanding of their GHG-science and the socio-economic issues involved in their adoption make them more long-term options. Practices like zero tillage, which continue to be adopted by prairie producers for economic and soil conservation reasons, could offer immediate GHG offsets without major economic incentives or changes in agricultural policy.

Canada will benefit from the adoption of these soil C sequestration practices even if soil sinks are not recognized internationally as valid removals of CO_2 from the atmosphere. Business-as-usual between 1990 and 2010, which included continued adoption of zero tillage practices, reduced frequency of summerfallow, and improved management of haylands and pastures, resulted in a reduction in emissions from soils from 6 Mt in 1990 to 0.5 Mt in 2010. Thus, even if the ~6 Mt sink in 2010 is not counted, the elimination of soils as a source of emissions could reduce agricultural GHG emissions BAU by about 6 Mt.

The mitigation scenarios indicate the magnitude of emission reductions, relative to the 2010 BAU and 1990 baseline scenarios, that could be achieved if the rate of adoption of sink-enhancing farming practices was accelerated above BAU. The results based on the Expert Opinion coefficients showed that increased zero tillage, reduced summerfallow frequency, conversion of poor-quality cropland to permanent cover, and improved management of grazing lands in total, reduce 2010 GHG emissions by 11.6 percent below 1990 levels. To rigorously assess the GHG reductions that could result from all of the strategies, a combined scenario would have to be developed and simulated. However, this 11.6 percent reduction indicates that C sequestration based on a combination of sink-enhancing practices could provide significant opportunities to offset agricultural GHG emissions.

The CEEMA-based modelling exercise demonstrates the importance of a systems approach to analysis of GHG. It is not possible to change one part of the system, such as acreage of zero tillage, the frequency of summerfallow, or the amount of permanent cover cropping, without cascading effects occuring. For example, although a reduction in summerfallow is associated with an increase in soil organic matter C and a removal of CO_2 from the atmosphere, it also increases GHG emissions in proportion to the increase in crop production. As a result, the net GHG mitigation potential from reduced summerfallow is lower than would be expected from estimates of only the sink potential. In fact, with the Century C sequestration coefficients there was a net increase in emissions from reduced summerfallow relative to the 2010 BAU scenario.

Because of this systems approach, the CEEMA analyses are informative but the results must be interpreted with recognition of the large degree of uncertainty associated with many coefficients and assumptions concerning adoption rates. The differences observed between the two sets of C sequestration coefficients indicate the level of scientific uncertainty. The GHG coefficients are essentially indicators of emission sources or sinks, and provide information about trend changes in emissions levels in response to changes in farming practices and levels of agricultural activity. Although the scenario analyses provide the direction of trend changes and the relative magnitude of emissions compared to the baseline conditions, they are only predictions based on estimates. At this early stage in the research, the uncertainty in GHG emission predictions must be taken into account when planning reduction strategies.

The ancillary benefits of increased soil C should also be considered as GHG reduction strategies for agriculture are developed. Sequestered C represents not only a reduction in atmospheric CO_2, but also an increase in soil organic matter, considered an important indicator of soil quality for crop production (Doran and Parkin 1994; Boehm and Anderson 1997). Further agricultural policies that encourage reduced net GHG emissions through soil sinks as well as increased soil C would be in keeping with Canada's policy of sustainable land management, since the adoption of zero tillage, the elimination of summerfallow, and the conversion of marginal land to permanent cover crops have been recognized as key factors in the trend toward improved soil quality and sustainability (Acton and Gregorich 1995) in Canadian agriculture. Even if these practices are not accepted as allowable removals of CO_2 internationally, their ability to remove CO_2 from the atmosphere should become part of domestic agricultural policy.

Even with the uncertainty, the options for achieving significant reductions in GHG emissions from alternate agricultural practices are limited, thus sink-enhancing practices offer the best short-term option for emission reductions from agriculture. Further, they provide other environmental and economic benefits besides GHG mitigation, promoting agricultural sustainability. We recognise that C sequestration will not solve the problem of GHG emissions for Canada or for agriculture, however, it is useful for the removal of atmospheric CO_2 while other technologies for GHG reduction can be developed in an effort to move society toward an economy less reliant on fossil fuel.

Acknowledgments
The authors would like to thank O. Bussler, C. Dauncey, R. Gill and S. Weseen for their assistance in completing the analysis and in the development of the base model.

CONTRIBUTORS

PIER L. BINDA is Professor of Geology at the University of Regina. His research interests span from stratigraphy to mineral deposits. He is currently co-leader of a UNESCO-IGCP project in Central Africa.

Pier L. Binda, Department of Geology, University of Regina, Regina, SK S4S 0A2, email: plbinda@sk.sympatico.ca

MARIE BOEHM is a soil scientist with Research Branch and Policy Branch of Agriculture and Agri-Food Canada. She is currently involved with efforts to estimate greenhouse gas emissions from Canadian agriculture and the development of models to test greenhouse gas mitigation and adaptation strategies.

M.M. Boehm, Centre for Agriculture, Law and Environment, University of Saskatchewan, Saskatoon, SK, 51 Campus Drive, S7N 5A8

MATTHEW BOYD holds a Ph.D. in Archaeology from the University of Calgary. His research interests are focused on the use and manipulation of pre-settlement landscapes by hunter-gatherers in North America. He is currently a postdoctoral researcher in the Department of Geological Sciences at the University of Manitoba.

Matthew Boyd, Department of Archaeology, University of Calgary, 2500 University Dr. N.W., Calgary, AB T2N 1N4, (403) 220-6956, FAX: (403) 282-9567, email: mjboyd@ucalgary.ca

KAREN DE BRES received her bachelor's and master's degrees from the University of Missouri-Columbia and a doctorate from Columbia University in New York. She has been a member of the Geography Department of Kansas State University since 1990, where she is currently an Associate Professor. Agreeing with Walt Whitman that the prairies and plains are North America's most characteristic landscape, her research interests include cultural geography, gender and tourism.

Karen DeBres, Department of Geography, Kansas State University, Manhattan, KS 66502, email: karendb@ksu.edu

RAYMOND L. DESJARDINS is one of Canada's experts with respect to understanding the dynamics and quantification of greenhouse gases in the environment. He has developed unique micrometeorological technology for measuring mass and energy fluxes at the field scale, extending these measurements to large-scale experiments using aircraft and satellite systems and integrating the results with other biological and physical information to quantify greenhouse gas exchange at a regional scale. He is currently a research scientist in the Research Branch of Agriculture and Agri-Food Canada and in charge of a national research project on the enhancement of the Agricultural greenhouse gas sinks.

R.L. Desjardins, Research Branch, Agriculture and Agri-Food Canada, Ottawa, ON, K1A 0C6

PATRICK C. DOUAUD is Associate Professor of Education at the University of Regina and editor-in-chief of *Prairie Forum*. His research interests include the history of ideas and alternative medicine.

Patrick C. Douaud, Canadian Plains Research Center, University of Regina, Regina, SK S4S 0A2, email: patrick.douaud@uregina.ca

TREVOR HARRISON is an instructor in Sociology at the University of Alberta. His areas of research include Canadian society, political sociology, and public policy. He is the author of *Of Passionate Intensity: Right-Wing Populism and the Reform Party of Canada* (1995). He is also co-editor of *The Trojan Horse: Alberta and the Future of Canada* (1995); and *Contested Classrooms: Education, Globalization, and Democracy in Alberta* (1999).

Trevor Harrison, Department of Sociology, University of Alberta, Edmonton, AB T6G 2H4 email: harrison@gpu.srv.ualberta.ca

David Hopkins is an Assistant Professor of Soil Science at North Dakota State University. He has worked in soil genesis and classification for the North Dakota Agricultural Experiment Station since 1981 and teaches the introductory soils course. His research interests include soil geography, soil spatial variability, and soil genesis.

David Hopkins, Department of Soil Science, 225 Walster Hall, North Dakota State University, Fargo, ND 58105

BRUCE JUNKINS is a senior economist with the Strategic Policy Branch of Agriculture and Agri-Food Canada. He has 25 years of experience, and recently has been involved with multi-disciplinary teams in developing integrated models to simultaneously estimate both the economic and environmental impacts of agricultural policies.

B. Junkins, Economic and Policy Analysis Directorate, Policy Branch, Agriculture and Agri-Food Canada, Ottawa, ON, K1A 0C6

SUREN KULSHRESHTHA is currently a Professor of Agricultural Economics, at the University of Saskatchewan. His current area of interest is greenhouse gas emissions from agriculture. He has recently served on the federal Agriculture and Agri-Food Issue Table. His other areas of interest include agroforestry, economic valuation of natural resources, and economic impact analysis.

S. Kulshreshtha, Agricultural Economics, University of Saskatchewan, 51 Campus Drive, Saskatoon, SK, S7N 5A8

VASU NAMBURIRI is a paleobotanist specialized in Cretaceous-Tertiary vegetation of North America and India. He has published articles and book chapters in a variety of international journals and symposium volumes.

E.M. Vasu Nambudiri, Sir Wilfred Grenfell College, Memorial University of Newfoundland, Corner Brook, NF A2H 6P9

ROSE OLFERT is Associate Professor in the Department of Agricultural Economics, University of Saskatchewan, Saskatoon, Saskatchewan. She has published in the areas of rural development, off-farm employment, farm women, restructuring of the rural economy, multipliers, and spatial labour market areas.

Rose Olfert, Department of Agricultural Economics, #3D34 Agriculture Building, 51 Campus Drive, University of Saskatchewan, Saskatoon, SK S7N 5A8 olfertr@duke.usask.ca

ALEC PAUL is Professor of Geography at the University of Regina. His research interests include the geography of western Canada, the climatology of prairie hailstorms and tornadoes, and railway landscapes.

Alec Paul, Department of Geography, University of Regina, SK S4S 0A2, email: paula1@meena.cc.uregina.ca

DUANE PELTZER is a plant ecologist and recently received his Ph.D. from the University of Regina. He is now doing postdoctoral work in New Zealand.

Duane A. Peltzer, Landcare Research, P.O. Box 69, Lincoln 8152, New Zealand; e-mail: PeltzerD@landcare.cri.nz

TODD A. RADENBAUGH is a Visiting Assistant Professor at The George Washington University and recently finished his Ph.D. at the University of Regina. In addition to his interest in interdisciplinary research spanning paleoecology and geography, he also studies human influences on ecosystem level processes.

Todd A. Radenbaugh, Canadian Plains Research Center, University of Regina, Regina, SK S4S 0A2, email: toddr@cas.uregina.ca

GARRY RUNNING is Assistant Professor of Geography at the University of Wisconsin-Eau Claire. He teaches two courses on physical geography and has two upper level courses on soils and soil geomorphology. Dr. Running has ongoing research in dunefields near Brandon, Manitoba. Both authors conducted their dissertation research in the North Dakota Sandhills.

Garry L. Running IV, Department of Geography, University of Wisconsin-Eau Claire, Eau Claire, WI, 54702-4004

ALAN R. SMITH has had a lifelong interest in the ornithology and geography of Saskatchewan. In addition to being the author of the comprehensive *Atlas of Saskatchewan Birds*, his research efforts for the Canadian Wildlife service involve travelling the province for new bird sighting and records, and banding song birds at the Last Mountain Bird Observatory (LMBO). He is also the provincial coordinator of the annual Breeding Bird Survey (BBS).

Alan R. Smith, Canadian Wildlife Service, Environment Canada; email: Alan.Smith@ec.gc.ca

JACK STABLER is a Professor in the Department of Agricultural Economics, University of Saskatchewan. His research and publications are in the areas of regional and rural development in western Canada and western United States.

Jack Stabler, Department of Agricultural Economics, #3D34 Agriculture Building, 51 Campus Drive, University of Saskatchewan, Saskatoon, SK S7N 5A8 stefaniu@duke.usask.ca

LITERATURE CITED

Introduction

Aberson, J.L. 1991. *From the Prairies with Hope*, R.E. Vandervennen ed. Regina: Canadian Plains Research Center.

Basso, K. 1996. *Wisdom Sits in Places: Landscape and Language Among the Western Apache.* Albuquerque: University of New Mexico Press.

Botsford, L.W., J.C. Castilla, and C.H. Peterson. 1997. "The Management of Fisheries and Marine Ecosystems." *Science* 277: 509–15.

Collins, S.L. 1992. "Fire Frequency and Community Heterogeneity in Tallgrass Prairie Vegetation." *Ecology* 73: 2001–06.

Collins, S.L. and L.L. Wallace. 1990. *Fire in North American Tallgrass Prairies.* Norman: University of Oklahoma Press.

Coues, S.L. 1878. "Field-notes on Birds Observed in Dakota and Montana Along the 40th Parallel During the Seasons of 1873 and 1874." *Bulletin of the U.S. Geological and Geographical Survey of the Territories* 4: 545–661.

Dyson, I.W. 1996. "Canada's Prairie Conservation Action Plan." In: Samson, F.B. and F.L. Knopf eds. *Prairie Conservation: Preserving North America's Most Endangered Ecosystem.* Washington, DC: Island Press, 175–86.

Frank, D.A., S.J. McNaughton, and B.F. Tracy 1998. "The Ecology of the Earth's Grazing Ecosystems." *Bioscience* 48: 513–21.

Freisen, G. 1984. *The Canadian Prairies: A History.* Toronto: University of Toronto Press.

Gould, S.J. 1989. *Wonderful Life: The Burgess Shale and the Nature of History.* New York: W.W. Norton and Company.

Grime, J.P. 1997. "Biodiversity and Ecosystem Function: The Debate Deepens." *Science* 277: 1260–61.

Hooper, D.P and P.M. Vitousek. 1997. "The Effects of Plant Composition and Diversity on Ecosystem Processes." *Science* 277: 1302–5.

Hughes, L. 2000. "Biological Consequences of Global Warming: Is the Signal Already Apparent?" *Trends in Ecology and Evolution* 15: 56–61.

Hughes, T.P. 1994. "Catastrophes, Phase Shifts, and Large-scale Degradation of a Caribbean Coral Reef." *Science* 265: 1547–51.

Hutchinson, G.E. 1965. *The Ecological Theater and the Evolutionary Play.* New Haven: Yale University Press.

Jackson, J.B.C. 1994. "Community Unity?" *Science* 264: 1412–13.

Kaiser, J. and R. Gallagher. 1997. "How Humans and Nature Influence Ecosystems." *Science* 277:1204–5.

Knapp, A.K., J.M. Blair, J.M. Briggs, S.L. Collins, D.C. Hartnett, L.C. Johnson, and E.G. Towne. 1999. "The Keystone Role of Bison in North American Tallgrass Prairie." *Bioscience* 48: 39–50.

Lapointe, B.E. and J. O'Connell. 1989. "Nutrient-enhanced Growth of *Cladophora prolifera* in Harrington Sound, Bermuda: Eutrophication of a Confined, Phosphorus-limiting Marine System." *Estuarine, Coastal, and Shelf Science* 28: 347–60.

Lawton, J.H. 1987. "Are There Assembly Rules for Successional Communities?" In A.J. Gray, M.J Crawely, and P.J. Edwards, eds. *Colonization, Succession, Stability.* Oxford: Blackwell Scientific Press, 225–45.

Mainguet, M. 1994. *Desertification: Natural Background and Human Mismanagement.* New York: Springer-Verlag.

McMichael, A.J. 1997. "Global Environmental Change and Human Health: Impact Assessment, Population Vulnerability, and Research Priorities." *Ecosystem Health* 3: 200-210.

Naeem, S.L., J. Thompson, S.P. Lawler, J.H. Lawton, and R.M. Woodfin. 1994. "Declining Biodiversity Can Alter the Performance of Ecosystems." *Nature* 368: 734–37.

Nee, S. and R.M. May. 1997. "Extinction and the Loss of Evolutionary History." *Science* 278: 692–93.

Reynaud, A. 1971. *Épistémologie de la géomorphologie.* Paris: Masson.

Rosemond, A.D. 1996. "Indirect Effects of Herbivores Modify Predicted Effects of Resources and Consumption of Plant Biomass." In: G.A. Polis and K.O. Winemiller eds. *Food Webs: Integration of Pattern and Dynamics.* New York: Chapman and Hall, 149–59.

Rowe, J.S. 1969. "Lighting Fires in Saskatchewan Grasslands." *Canadian Field Naturalist* 83: 317–24.

Samson, F.B. and F.L. Knopf eds. 1996. *Prairie Conservation: Preserving North America's Most Endangered Ecosystem.* Washington, DC: Island Press.

Schindler, D.W. 1998. "A Dim Future for Boreal Waters and Landscapes: Cumulative Effects of Climatic Warming, Stratospheric Ozone Depletion, Acid Precipitation and Other Human Activities." *BioScience* 48: 157–64.

Schlesinger, W.H., J.F. Reynolds, G.L. Cunningham, L.F. Huenneke, W.M. Jarrell, R.A. Virginia, and W.G. Whitford. 1990. "Biological Feedbacks in Global Desertification." *Science* 247: 1043–48.

Seastedt, T. R. 1995. "Soil Systems and Nutrient Cycles of the North American Prairie." In Joern, A. and K.H. Keeler eds. *The Changing Prairie: North American Grasslands.* New York: Oxford University Press, 157–74.

Selby, C.J. and M.J. Santry. 1996. *A National Ecological Framework for Canada: Data Model, Database and Programs.* Ottawa/Hull: Agriculture and Agri-Food Canada.

Simpson, R.D. and N.L. Christensen. 1997. *Ecosystem Function and Human Activities: Reconciling Economics and Ecology.* New York: Chapman and Hall.

Symstad, A., J.D. Tilman, J. Willson, and J.M.H. Knops. 1998. "Species Loss and Ecosystem Functioning: Effects of Species Identity and Community Composition." *Oikos* 81: 389–97.

Tilman, D.J. 1996. "Biodiversity: Populations Versus Ecosystem Stability." *Ecology* 77: 350–63.

Tilman, D., J. Knops, D. Wedin, P. Reich, M. Ritchie, and E. Siemann. 1997. "The Influence of Functional Diversity and Composition on Ecosystem Processes." *Science* 277: 1300-1302.

Valentine, J.W. and D. Jablonski. 1993. "Fossil Communities: Compositional Variation at Many Time Scales." In: Ricklefs, R.E. and D. Schluter eds. *Species Diversity in Ecological Communities: Historical and Geographical Perspectives.* Chicago: University of Chicago Press, 341–49.

Vitousek, P.M., H.A. Mooney, J. Lubchenco, and J.M. Melillo. 1997. "Human Domination of the Earth's Ecosystems." *Science* 277: 494–99.

Wright, H.A., and A.W. Bailey 1980. *Fire Ecology and Prescribed Burning in the Great Plains.* General Technical Report INT-77. Intermountain Forest and Range Experiment Station, U.S. Department of Agriculture USDA Forest Service.

Yovel, Y. 1980. *Kant and the Philosophy of History.* Princeton: Princeton University Press.

Zochert, D. 1976. *Laura: The Life of Laura Ingalls Wilder.* Chicago: H. Regnery Co.

Chapter 1

Allan, J.A. and J.O.C. Sanderson. 1945. *Geology of the Red Deer and Rosebud Sheets, Alberta.* Report 13. Edmonton: Research Council of Alberta.

Bennett, M.R. and P. Doyle. 1996. "Global cooling inferred from dropstones in the Cretaceous: fact or wishful thinking?" *Terra Nova* 74: 182–85.

Binda, P.L. 1970. "Sedimentology and Vegetal Micropaleontology of the Rocks Associated with the Kneehills Tuff of Alberta." Ph.D. dissertation, University of Alberta.

——. 1992. "The Battle Formation: A Lacustrine Episode in the Late Maastrichtian of Western Canada." In N.J. Mateer and Pei-Ji Chen, eds., *Aspects of Non-marine Cretaceous Geology.* Beijing: China Ocean Press.

Binda, P.L., E.M.V. Nambudiri, S.K. Srivastava, M. Schmitz, A. Longinelli and P. Iacumin. 1991. "Stratigraphy, Paleontology, and Aspects of Diagenesis of the Whitemud Formation (Maastrichtian) of Alberta and Saskatchewan." In J.E. Christopher and F.M. Haidl eds. *Sixth International Williston Basin Symposium.* Regina: Saskatchewan Geological Society, 179–92.

Binda, P.L. and B.R. Watters. 1997. "Pyroclastic Quartz Grains from the Late Cretaceous Battle Formation of Southwest Saskatchewan." Miscellaneous Report 97–4. Regina: Saskatchewan Geological Survey, Saskatchewan Energy and Mines, 213–15.

Caldwell, W.G.E. 1982. "The Cretaceous System in the Williston Basin – A Modern Appraisal." In J.E. Christopher and D.M. Kent eds. *Fourth International Williston Basin Symposium.* Regina: Saskatchewan Geological Society.

——. 1984. "Early Cretaceous Transgressions and Regressions in the Southern Interior Plains." In D.F. Stott and D.J. Glass eds. *The Mesozoic of Middle North America.* Calgary: Canadian Society of Petroleum Geologists.

Folinsbee, R.E., H. Baadsgaard, G.L. Cumming, J. Nascimbene and M. Shafiqullah. 1965. "Late Cretaceous Radiometric Dates from the Cypress Hills of Western Canada." *15th Annual Field Conference Guidebook, Part I, Cypress Hills Plateau.* Edmonton: Alberta Society of Petroleum Geologists, 162–74.

Frank, M.C. 1999. "Organic Petrology and Depositional Environment of the Souris Lignite, Ravenscrag Formation (Paleocene), Southern Saskatchewan, Canada." Ph.D. dissertation, University of Regina.

Fyfe, W.S. 1992. "Global Change: Anthropologic Forcing – The Moving Target." *Terra Nova* 4: 284–87.

Hallam, A. 1981. *Facies and the Stratigraphic Record.* San Francisco: W.H. Freeman.

——. 1987. "End-Cretaceous Mass Extinction Event: Argument for Terrestrial Causation." *Science* 238: 1237–42.

——. 1993. "Jurassic Climates as Inferred from Sedimentary and Fossil Record." *Philosophical Transactions of the Royal Society of London,* 287–96.

Haq, B.U., J. Hardenbol and P.R. Vail. 1987. "Chronology of Fluctuating Sea Levels Since the Triassic." *Science* 235: 1156–68.

Jerzykiewicz, T. and A.R. Sweet. 1986. "Caliche and Associated Impoverished Palynological Assemblages: An Innovative Line of Paleoclimatic Research onto the Uppermost Cretaceous and Paleocene of Southern Alberta." Paper 86–1B. *Current Research, Part B, Geological Survey of Canada,* 653–63.

Kent, D.M. 1994. "Paleogeographic Evolution of the Cratonic Platform – Cambrian to Triassic." In G. Mossop and I. Shetsen comps. *Geological Atlas of the Western Canadian Sedimentary Basin.* Edmonton: Canadian Society of Petroleum Geologists and Alberta Research Council.

Lerbekmo, J.F., C. Singh, D.M. Jarzen and D.A. Russell. 1979. "The Cretaceous-Tertiary Boundary in South-central Alberta – A Revision Based on Additional Dinosaurian and Microfloral Evidence." Canadian Journal of Earth Sciences 16: 1866–69.

Matthews, R.K. 1984. *Dynamic Stratigraphy.* Englewood Cliffs: Prentice-Hall.

Nambudiri, E.M.V. and P.L. Binda. 1989. "Dicotyledonous Fruits Associated with Coprolites from the Upper Cretaceous (Maastrichtian) Whitemud Formation, Southern Saskatchewan, Canada." *Review of Palaeobotany and Palynology* 59: 57–66.

——. 1991. "Paleontology, Palynology and Depositional Environment of the Maastrichtian Whitemud Formation in Alberta and Saskatchewan, Canada." *Cretaceous Research* 12: 579–96.

Nambudiri, E.M.V., L.W. Vigrass and J.A. MacEchern. 1986. *Palynological Studies of Lower Cretaceous Strata from Saskatchewan in Canada.* Regina: University of Regina Energy Research Unit.

Obradovich, J.D. and W.A. Cobban. 1975. "A Time-scale for the Late Cretaceous of the Western Interior of North America." In W.G.E. Caldwell ed. *The Cretaceous System in the Western Interior of North America.* Waterloo: Geological Association of Canada, 31–54.

Potter, J., A.P. Beaton, W.J. McDougall, E.M.V. Nambudiri and L.W. Vigrass. 1991. "Depositional Environment of the Hart Coal Zone (Paleocene), Willowbunch Coalfield, Southern Saskatchewan, Canada from Petrographic, Palynological, Paleobotanical, Mineral and Trace Element Studies." *International Journal of Coal Geology* 19: 253–81.

Rahmani, R.A. 1981. *Facies Relationships and Paleoenvironments of a Late Cretaceous Tide-dominated Delta, Drumheller, Alberta. A Field Guide.* Edmonton: Edmonton Geological Society.

Ritchie, W.D. 1957. "The Kneehills Tuff." M.Sc. thesis, University of Alberta.

Russell, D.A. 1977. *A Vanished World. The Dinosaurs of Western Canada.* Ottawa: National Museum of Canada.

——. 1984. *A Check List of the Families and Genera of North American Dinosaurs.* Ottawa: National Museum of Canada.

Russell, L.S. 1983. "Evidence for an Unconformity at the Scollard-Battle Contact, Upper Cretaceous Strata, Alberta." *Canadian Journal of Earth Sciences* 20: 1219–31.

Schmitz. M. and P.L. Binda. 1991. "Coprolites from the Maastrichtian Whitemud Formation of Southern Saskatchewan: Morphological Classification and Interpretation on Diagenesis." *Paläontologische Zeitschrift* 66: 199–211.

Singh, C. 1964. *Microflora of the Lower Cretaceous Mannville Group, East-central Alberta.* Edmonton: Research Council of Alberta.

——. 1971. *Lower Cretaceous Microfloras of the Peace River Area, Northwestern Alberta.* Edmonton: Research Council of Alberta.

Srivastava, S.K. 1970. "Pollen Biostratigraphy and Paleoecology of the Edmonton Formation (Maastrichtian), Alberta, Canada." *Palaeogeography, Palaeoclimatology, Palaeoecology* 7: 221–76.

——. 1978. *Cretaceous Spore-Pollen Floras: A Global Evaluation.* Lucknow, India: International Publishers.

——. 1981. "Fossil Pollen Genus Kurtzipites Anderson." *Journal of Paleontology* 55: 868–79.

Srivastava, S.K. and P.L. Binda. 1984. "Siliceous and Silicified Microfossils from the Maastrichtian Battle Formation of Southern Alberta, Canada." *Paleobiologie Continentale* 14: 1–24.

Stanley, S.M. 1998. *Earth System History.* New York: W.H. Freeman.

Sweet, A.R., D.R. Braman and J.F. Lerbekmo. 1999. "Sequential Palynological Changes Across the Composite Cretaceous-Tertiary (K-T) Boundary Claystone and Contiguous Strata, Western Canada and Montana, U.S.A." *Canadian Journal of Earth Sciences* 36: 743–68.

Wall, J.W., A.R. Sweet and L.V. Hill. 1971. "Paleoecology of the Bearpaw and Contiguous Upper Cretaceous Formations in the C.P.O.G. Strathmore Well, Wouthern Alberta." *Bulletin of the Society of Petroleum Geologists* 19: 691–702.

Chapter 2

Adams, G. 1976. *Prehistoric Survey of the Lower Red Deer River 1976.* Edmonton: Archaeological Survey of Alberta.

Antevs, E. 1955. "Geologic-Climate Dating in the West." *American Antiquity* 20, no. 4: 317–35.

Barnosky, C.W. 1989. "Postglacial Vegetation and Climate in the Northwestern Great Plains." *Quaternary Research* 31: 57–73.

Beaudoin, A.B. 1992. "Early Holocene Paleoenvironmental Data Preserved in 'Non-Traditional' Sites." Regina: Palliser Triangle Global Change Meeting, Program and Abstracts.

Boldurian, A.T. 1991. "Folsom Mobility and Organization of Lithic Technology: A View from Blackwater Draw, New Mexico." *Plains Anthropologist* 36, no. 137: 281–95.

Boyd, M.J. 2000. "Late Quaternary Geoarchaeology of the Lauder Sandhills, Southwestern Manitoba, Canada." Ph.D. dissertation, University of Calgary.

Broilo, F.J. 1971. "An Investigation of Surface Collected Clovis, Folsom, and Midland Projectile Points from Blackwater Draw and Adjacent Localities." M.A. thesis, Eastern New Mexico University.

Brown, D.A. 1984. "Prospects and Limits of a Phytolith Key for Grasses in the Central United States." *Journal of Archaeological Science* 11: 345–68.

Brumley, J.H. 1975. *The Cactus Flower Site in Southeastern Alberta: 1972–1974 Excavations.* Ottawa: Archaeological Survey of Canada.

Buchner, A.P. 1982. *An Archaeological Survey of the Winnipeg River.* Winnipeg: Papers in Manitoba Archaeology.

Buchner, A.P. and L. Pettipas. 1990. "The Early Occupations of the Glacial Lake Agassiz Basin in Manitoba; 11,500 to 7,700 B.P." In N.P. Lasca and J. Donague eds. *Archaeological Geology of North America.* Boulder: Geological Society of America.

Buol, S.W., F.D. Hole and R.J. McCracken. 1989. *Soil Genesis and Classification.* Ames: Iowa State University Press.

Clayton, L. and S.R. Moran. 1982. "Chronology of Late Wisconsin Glaciation in Middle North America." *Quaternary Science Reviews* 1: 55–82.

COHMAP. 1988. "Climatic Changes of the Last 18,000 Years: Observations and Model Simulations." *Science* 241: 1043–52.

Deller, D.B., C.J. Ellis and L.T. Kenyon. 1986. "Archaeology of the Southeastern Huron Basin." In W. Fox ed. *Studies in Southwestern Ontario Archaeology.* London, ON: Ontario Archaeological Society. 1: 3–12.

Dormaar, J.F. and L.E. Lutwick. 1966. "A Biosequence of Soils of the Rough Fescue Prairie-Poplar Transition in Southwestern Alberta." *Canadian Journal of Earth Sciences* 3: 457–71.

Dyck, I. 1983. "The Prehistory of Southern Saskatchewan." In H.T. Epp and I. Dyck eds. *Tracking Ancient Hunters.* Saskatoon: Saskatchewan Archaeological Society.

Dyck, W., J.G. Fyles and W. Blake Jr. 1965. "Geological Survey of Canada Radiocarbon Dates IV." *Radiocarbon* 7: 24–46.

Eilers, R.G., L.A. Hopkins and R.E. Smith. 1978. *Soils of the Boissevain-Melita Area.* Winnipeg: Department of Agriculture.

Ellis, C.J. and D.B. Deller. 1990. "Paleo-Indians." In C.J. Ellis and N. Ferris eds. *The Archaeology of Southern Ontario to A.D. 1650.* London, ON: Ontario Archaeological Society. 5: 37–63.

Elson, J.A. 1983. "Glacial Lake Agassiz — Discovery and a Century of Research." In J.T. Teller and L. Clayton, eds. *Glacial Lake Agassiz.* Ottawa: Geological Association of Canada.

Environment Canada. 1993. *Canadian Climate Normals (Vol. 2).* Ottawa: Environment Canada, Atmospheric Environment Service.

Forbis, R.G. 1968. "Fletcher: a Paleo-Indian Site in Alberta." *American Antiquity* 33, no.1: 1–10.

Frison, G.C. 1971. "The Buffalo Pound in Northwestern Plains Prehistory: Site 48CA302." *Plains Anthropologist* 16, no. 54: 258–84.

Grimm, E.C. 1995. "Recent Palynological Studies from Lakes in the Dakotas." Lincoln, NE: Geological Society of America, North-Central Section-South-Central Section Meeting, Program and Abstracts.

Hamilton, S. and B.A. Nicholson. 1999. "Ecological Islands and Vickers Focus Adaptive Transitions in the Pre-contact Plains of Southwestern Manitoba." *Plains Anthropologist* 44, no. 167: 5–25.

Haraguchi, A. 1991. "Effect of a Flooding-Drawdown Cycle on Vegetation in a System of Floating Peat Mat and Pond." *Ecological Research* 6: 247–63.

Haug, J.K. 1976. The 1974–1975 *Excavations at the Cherry Point Site (DkMe-10):A Stratified Archaic Site in Southwest Manitoba.* Winnipeg: Papers in Manitoba Archaeology.

Haynes, C.V., Jr. 1964. "Fluted Projectile Points: Their Age and Dispersion." *Science* 145: 1408–13.

——. "Were Clovis Progenitors in Beringia?" In D.M. Hopkins, J.V. Matthews, Jr., C.E. Schweger and S.B. Young, eds. *Paleoecology of Beringia.* New York: Academic Press.

Hofman, J.L., D.S. Amick and R.O. Rose. 1990. "Shifting Sands: A Folsom-Midland Assemblage from a Campsite in Western Texas." *Plains Anthropologist* 33: 337–50.

Hofman, J.L. and R.W. Graham. 1998. "The Paleo-Indian Cultures of the Great Plains." In W.R. Wood, ed. *Archaeology on the Great Plains.* Lawrence: University Press of Kansas.

Hohn, S.L. and R.J. Parsons. 1993. *Lauder Sandhills Wildlife Management Area Natural Resources Inventory.* Winnipeg: Manitoba Natural Resources.

Jennings, J.D. 1955. *The Archaeology of the Plains: An Assessment (With Special Reference to the Missouri River Basin).* Salt Lake City: University of Utah.

Kelly, M.E. and B.E. Connell. 1978. *Survey and Excavations of The Pas Moraine: 1976 Field Season.* Winnipeg: Department of Tourism.

Knox, J.C. 1983. "Response of River Systems to Holocene Climates." In H.E. Wright, ed., *Late Quaternary Environments of the United States,* Volume 2. Minneapolis: University of Minnesota Press.

Largent, F.B., Jr., M.R. Waters and D.L. Carlson. 1991. "The Spatiotemporal Distribution and Characteristics of Folsom Projectile Points in Texas." *Plains Anthropologist* 36, no. 137: 323–41.

Lemmen, D.S. (ed.). 1996. *Landscapes of the Palliser Triangle, Guidebook for the Canadian Geomorphology Research Group Field Trip.* Saskatoon: Canadian Association of Geographers Annual Meeting.

Looman, J. and K.F. Best. 1987. *Budd's Flora of the Canadian Prairie Provinces.* Ottawa: Agriculture Canada.

Miller, S.J. and W. Dort, Jr. 1978. "Early Man at Owl Cave: Current Investigations at the Waden Site, Eastern Snake River Plains, Idaho." In A.L. Bryan, ed. *Early Man in America from a Circum-Pacific Perspective.* Edmonton: University of Alberta.

Nambudiri, E.M.V., J.T. Teller and W.M. Last. 1980. "Pre-Quaternary Microfossils — A Guide to Errors in Radiocarbon Dating." *Geology* 8: 123–26.

Nicholson, B.A. and S. Hamilton. 1998. "Middle Precontact Occupations at the Vera site in the Makotchi-Ded Dontipi Locale." Paper presented at the 1998 CAA Conference, Victoria, British Columbia.

Nielsen, E., E.M. Gryba and M.C. Wilson. 1984. "Bison Remains from a Lake Agassiz Spit Complex in the Swan River Valley, Manitoba; Depositional Environment and Paleoecological Implications." *Canadian Journal of Earth Sciences* 21, no. 7: 829–42.

Nielsen, E., K.D. McLeod, E. Pip and J.C. Doering. 1996. *Late Holocene Environmental Changes in Southern Manitoba (Field Trip A2).* Winnipeg: Geological Association of Canada.

Pettipas, L. 1967. "Paleo-Indian Manifestations in Manitoba: Their Spatial and Temporal Relationships with the Campbell Strandline." M.A. thesis, University of Manitoba.

Pettipas, L. and A.P. Buchner. 1983. "Paleoindian Prehistory of the Glacial Lake Agassiz Region in Manitoba, 11,500 to 6500 BP." In J.T. Teller and L. Clayton, eds. *Glacial Lake Agassiz.* Ottawa: Geological Association of Canada.

Quigg, J.M. 1976. "A Note on the Fletcher Site." In J.M. Quigg and W.J. Byrne, eds. *Archaeology in Alberta, 1975.* Edmonton: Archaeological Survey of Alberta.

Ritchie, J.C. 1969. "Absolute Pollen Frequencies and Carbon-14 Age of a Section of Holocene Lake Sediment from the Riding Mountain Area of Manitoba." *Canadian Journal of Botany* 47: 1345–49.

——. 1976. "The Late-Quaternary Vegetational History of the Western Interior of Canada." *Canadian Journal of Botany* 54: 1793–1818.

Ritchie, J.C. and S. Lichti-Federovich. 1968. "Holocene Pollen Assemblages from the Tiger Hills, Manitoba." *Canadian Journal of Earth Sciences* 5: 873–80.

Running IV, G.L. 1995. "Archaeological Geology of the Rustad Quarry Site (32RI775): an Early Archaic Site in Southeastern North Dakota." *Geoarchaeology: An International Journal* 10, no. 3: 183–204.

Scoggan, H.J. 1953. "Botanical Investigations in the Glacial Lakes Agassiz-Souris Basins, 1951." In *Annual Report of the National Museum of Canada for the Fiscal Year 1951–1952.* Ottawa: n.p.

Sellards, E., G. Evans and G. Mead. 1947. "Fossil Bison and Associated Artifacts from Plainview, Texas, With Description of Artifacts by Alex D. Krieger." *Geological Society of America Bulletin* 58: 927–54.

Shang, Y. and W.M. Last. 1999. "Mineralogy, Lithostratigraphy, and Inferred Geochemical History of North Ingebrigt Lake, Saskatchewan." In D.S. Lemmen and R.E. Vance, eds. *Holocene Climate and*

Environmental Change in the Palliser Triangle: A Geoscientific Context for Evaluating the Impacts of Climate Change on the Southern Canadian Prairies. Ottawa: Geological Survey of Canada.

Soil Classification Working Group 1998. *The Canadian System of Soil Classification.* Ottawa: Agriculture and Agriculture-Food Canada.

Sorenson, C.J. 1977. "Reconstructed Holocene Bioclimates." *Annals of the Association of American Geographers* 67, no. 2: 214–22.

Steinbring, J. 1970. "Evidences of Old Copper in a Northern Transitional Zone." In W.M. Hlady, ed. *Ten Thousand Years: Archaeology in Manitoba.* Winnipeg: Manitoba Archaeological Society.

Stewart, A. 1984. "The Zander Site: Paleo-Indian Occupation of the Southern Holland Marsh Region of Ontario." *Ontario Archaeology* 41: 45–79.

Storck, P.L. 1982. "Paleo-Indian Settlement Patterns Associated with the Strandline of Glacial Lake Algonquin in Southcentral Ontario." *Canadian Journal of Archaeology* 6: 1–31.

Stuiver, M., and P.J. Reimer. 1993. "Extended 14C Data Base and Revised Calib 3.0 ^{14}C Age Calibration Program." *Radiocarbon* 35, no. 1: 215–30.

Sun, C. 1996. "Sedimentology and Geomorphology of the Glacial Lake Hind Area, Southwestern Manitoba, Canada." Ph.D. dissertation, University of Manitoba.

Sun, C., and Teller, J.T. 1997. "Reconstruction of Glacial Lake Hind in Southwestern Manitoba, Canada." *Journal of Paleolimnology* 17: 9–21.

Teller, J.T., L.H. Thorleifson, G. Matile and W.C. Brisbin. 1996. S*edimentology, Geomorphology and History of the Central Lake Agassiz Basin.* Winnipeg: Geological Association of Canada/Mineralogical Association of Canada.

Twiss, P.C. 1992. "Predicted World Distribution of C3 and C4 Grass Phytoliths." In G. Rapp Jr. and S.C. Mulholland, eds. *Phytolith Systematics.* New York: Plenum Press.

Vance, R.E. 1997. "The Geological Survey of Canada's Palliser Triangle Global Change Project: A Multidisciplinary Geolimnological Approach to Predicting Potential Global Change Impacts on the Northern Great Plains." *Journal of Paleolimnology* 17: 3–8.

Vickers, J.R. 1986. *Alberta Plains Prehistory: A Review.* Edmonton: Archaeological Survey of Alberta. 27: 1–139.

Vickers, J.R. and A. Beaudoin. 1989. "A Limiting AMS Date for the Cody Complex Occupation at the Fletcher Site, Alberta." *Plains Anthropologist* 34: 261–64.

Warren, T. 1980. "The Ancient Cultural Developments in Southwestern Manitoba as Seen in the Gould and Warren Collections." *Archae-Facts* 7, no. 2: 6–21.

Wendorf, F., A.D. Krieger, C.C. Albritton and T.D. Stewart. 1955. *The Midland Discovery.* Austin: University of Texas Press.

Wettlaufer, B.N. and W.J. Mayer-Oakes. 1960. *The Long Creek Site.* Regina: Saskatchewan Department of Natural Resources.

Wheat, J. 1972. "The Olsen-Chubbuck Site; A Paleo-Indian Bison Kill." *Society for American Archaeology Memoir* 26.

Wipff, J.K. 1996. Nomenclatural Combinations in the *Andropogon gerardii* Complex (Poaceae: Andropogoneae)." *Phytologia* 80, no. 5: 343–47.

Wormington, H.M. and R.G. Forbis. 1964. *An Introduction to the Archaeology of Alberta, Canada.* Colorado: Denver Museum of Natural History.

Wright, H.E., J.E. Kutzbach, T. Webb III, W.F. Ruddiman, F.A. Street-Perrott and P. J. Bartlein eds. 1993. *Global Climates Since the Last Glacial Maximum.* Minneapolis: University of Minnesota.

Wright, J.V. 1995. *A History of the Native Peoples of Canada,* Vol. 1. Hull: Canadian Museum of Civilization.

Xia, J., B.J. Haskell, D.R. Engstrom, and E. Ito. 1997. "Holocene Climate Reconstructions from Tandem Trace-Element and Stable Isotope Composition of Ostracodes from Coldwater Lake, North Dakota, USA." *Journal of Paleolimnology* 17: 85–100.

Yansa, C.H. and J.F. Basinger. 1999. "A Postglacial Plant Macrofossil Record of Vegetation and Climate Change in Southern Saskatchewan." In D.S. Lemmen and R.E. Vance, eds. *Holocene Climate and Environmental Change in the Palliser Triangle: A Geoscientific Context for Evaluating the Impacts of Climate Change on the Southern Canadian Prairies.* Ottawa: Geological Survey of Canada.

Chapter 3

Armstrong. C.A. 1982. *Ground-water Resources of Ransom and Sargent Counties, North Dakota.* Bismarck: North Dakota State Water Commission.

Baker, C.H. 1967. "New observations on the Sheyenne delta of Glacial Lake Agassiz." *U.S. Geological Survey Professional Paper* 575-B, 62-68.

Barbour, M.G., J.H. Burk, and W.D. Pitts. 1987. *Terrestrial Plant Ecology (2nd ed.).* Menlo Park, CA.: Benjamin/Cummings.

Barker, W.T., and W.C. Whitman. 1989. *Vegetation of the Northern Great Plains.* Fargo: North Dakota Agricultural Extension Service.

Bell, W.B. 1910. "Observations and Recommendations Based Upon the Biological Survey at McLeod, Sandoun Twp., Ransom County, North Dakota." In *Fourth Biennial Report of the Agricultural College Survey of North Dakota For the Years 1907 and 1908.* Fargo: North Dakota State University.

Bluemle, J.P. 1979. *Geology of Ransom and Sargent Counties, North Dakota.* Bismarck: North Dakota Geological Survey.

Brady, N.C. 1990. *The Nature and Properties of Soils. 10th ed.* New York: Macmillan.

Brophy, J.A. 1967. "Some Aspects of the Geological Deposits of the South End of the Lake Agassiz Basin." In W. J. Mayer-Oakes ed. *Life, Land and Water.* Winnipeg: University of Manitoba Press.

Brophy, J.A., and J.P. Bluemle. 1983. "The Sheyenne River: Its Geological History and Effects on Lake Agassiz." In J.T. Teller and L. Clayton eds. *Glacial Lake Agassiz.* Toronto: Geological Association of Canada.

Burgess, R.L. 1965. "A Study of Plant Succession in the Sandhills of Southeastern North Dakota." *Proceedings of the North Dakota Academy of Science* 19: 62–80.

Clayton, L., S.R. Moran, and B.W. Bickley, Jr., 1976. *Stratigraphy, Origin, and Climatic Implications of Late-Pleistocene Upland Silts in North Dakota.* Bismarck: North Dakota Geological Survey.

Danbom, D.B. 1990. *Our Purpose is to Serve: The First Century of the North Dakota Agricultural Experiment Station.* Fargo: North Dakota State University.

David, P.P. 1971. "The Brookdale Road Section and Its Significance in the Chronological Studies of Dune Activities in the Brandon Sand Hills of Manitoba." In A.C. Turnock, ed. *Geoscience Studies in Manitoba.* Toronto: Geological Association of Canada.

——. 1977. *Sand Dune Occurrences of Canada.* Ottawa: Indian and Northern Affairs.

——. 1979. "Sand dunes in Canada." *Geos* (Spring): 12-14.

Ely, C.W., R.E. Willard, and J.T. Weaver. 1907. *Soil Survey of Ransom County, North Dakota.* Washington, DC: United States Department of Agriculture.

Great Plains Flora Association. 1986. *Flora of the Great Plains.* Lawrence: University of Kansas Press.

Harris, K.L. 1987. *Surface Geology of the Sheyenne River Map Area: AS-15-A1, 1: 250,000.* Bismarck: North Dakota Geological Survey.

Hopkins, D.G. 1997. "Hydrologic and Abiotic Constraints on Soil Genesis and Natural Vegetation Patterns in the Sandhills of North Dakota." Ph.D. dissertation, North Dakota State University.

Katz, R.W., and Brown, B.G. 1992. "Extreme Events in a Changing Climate: Variability is More Important Than Averages." *Climate Change* 21: 289-309.

Laird, K.R., S.C. Fritz, K.A. Maasch, and B. F. Cumming. 1996. "Greater Drought Intensity and Frequency Before AD 1200 in the Northern Great Plains, USA." *Nature* 384: 552-54.

Last, W.M., and Vance, R.E. 1998. *Holocene Climate Variability on the Southern Interior Plains, Canada; Limnological Reconstructions; The Conversation.* Boulder, CO: Geological Society of America.

Lemmen, D.S. 1996. *Landscapes of the Palliser Triangle, Guidebook for the Canadian Geomorphology Research Group Field Trip.* Saskatoon: University of Saskatchewan.

Lemmen, D.S., R.E. Vance, I.A. Campbell, D.J. Pennock, P.P. David, D.J. Sauchyn, and S.A. Wolfe. 1998. "Geomorphic Systems of the Palliser Triangle, Southern Canadian Prairies; Description and Response to Changing Climate." In *Bulletin of the Geological Survey of Canada.*

McKinstry, H.C. 1910. "An Investigation of the Soils in the Vicinity of McLeod, North Dakota." In *Fifth Biennial Report of the Agricultural College Survey of North Dakota for the Years 1909 and 1910.* Fargo: North Dakota State University.

McNaughton, S.J., M.B. Coughenour, and L.L. Wallace. 1974. "Interactive Processes In Grassland Ecosystems." In J.R Estes, R.J. Tyrl, and J.N. Brunken eds., *Grasses and Grasslands Systematics and Ecology.* Norman: University of Oklahoma Press.

Manske, L.L. 1980. "Habitat, Phenology and Growth of Selected Sandhills Range Plants." Ph. D. dissertation, North Dakota State University.

Manske, L.L., W.T. Barker, and M.E. Biondini. 1988. "Effects of Grazing Management on Grassland Plant Communities and Prairie Grouse Habitat." In *Prairie Chickens on the Sheyenne National Grasslands.* Ft. Collins, CO: United States Department of Agriculture.

Muhs, D.R., T.W. Stafford, Jr., J. Been, S.H. Mahon, J. Burdett, G. Skipp, and Z. Muhs Rowland. 1997. "Holocene Eolian Activity in the Minot Dune Field, North Dakota." *Canadian Journal of Earth Science* 34: 1442–59.

NCR-13. 1998. *Recommended Chemical Soil Test Procedures for the North Central Region.* Columbia: Missouri Agricultural Experiment Station.

Nelson, W.T. 1986. "Grassland Habitat Type Classification of the Sheyenne National Grasslands of Southeastern North Dakota." M.S. thesis, North Dakota State University.

North Dakota Agricultural Statistics Service. 1999. *North Dakota Agricultural Statistics 1999.* Fargo: United States Department of Agriculture.

Robinson, E.B. 1966. *History of North Dakota.* Lincoln: University of Nebraska Press.

Running IV, G.L., 1996. "The Sheyenne Delta From the Cass Phase to the Present - Landscape Evolution and Paleoenvironment." In K.L. Harris, M.R. Luther, and J.R. Reid eds., *Quaternary Geology of the Southern Lake Agassiz Basin.* Bismarck: North Dakota Geological Survey, Miscellaneous Series 82.

———. 1997. "Geomorphology, Stratigraphy, and Landscape Evolution on the Sheyenne Delta, Southeastern North Dakota; Implications for Holocene Paleoenvironmental Change on the Northeastern Great Plains." Ph. D. dissertation, University of Wisconsin-Madison.

Running, IV, G.L., and T. Boutton. 1996. "Late-Holocene Parabolic Dunes on the Sheyenne Delta, Southeastern North Dakota." Saskatoon: Canadian Association of Geographers Annual Meeting, Geological Survey of Canada Special Aeolian Session.

Seiler, G.J. and W.T. Barker. 1985. "Vascular flora of Ransom, Richland, and Sargent Counties, North Dakota." *Prairie Naturalist* 17, no. 4: 193–240.

Shunk, R.A. 1917. "Plant associations of Shenford and Owego Townships, Ransom County, North Dakota." M.S. thesis, University of North Dakota.

Terzhagi, K. and R.B. Peck. 1948. *Soil Mechanics in Engineering Practice.* New York: John Wiley and Sons.

Thorfinnson, S.M. 1975. *Ransom County History.* Lisbon: Ransom County Historical Society.

USFS. 1980. "*Sheyenne National Grassland Land Management Plan, Richland and Ransom Counties, North Dakota.*" Billings, MT: United States Department of Agriculture.

Van Dyne, G. 1979. "The Nature of Grassland." In R.T. Coupland ed. *Grassland Ecosystems of the World: Analysis of Grasslands and Their Uses.* Cambridge: Cambridge University Press.

Vance, R.E., and W.M. Last. 1994. "Paleolimnology and Global Change on the Southern Canadian Prairies." In *Current Research: Interior Plains and Arctic Canada—Plaines interieures et region arctique du Canada.* Ottawa: Geological Survey of Canada.

Wali, M.K., G.W. Dewald, and S M. Jalal. 1973. "Ecological Aspects of Some Bluestem Communities in the Red River Valley." *Bulletin of the Torrey Botanical Club* 100: 339–48.

Wanek, W.J. 1964. "The Grassland Vegetation of the Sandhills Region of Southeastern North Dakota." M.S. thesis, North Dakota State University.

Wolfe, S.A. D.J., Huntley, and J. Ollerhead. 1994. "Recent and Late Holocene Sand Dune Activity in Southwestern Saskatchewan." In *Current Research 1995-B.* Ottawa: Geological Survey of Canada.

Chapter 4

Agriculture Canada. 1992. *Soil Landscapes of Canada: Saskatchewan.* Ottawa: Agriculture Canada.

Archibold, O.W. 1995. *Ecology of World Vegetation.* London: Chapman and Hall.

Axelrod, D.I. 1985. "Rise of the Grassland Biome, Central North America." *The Botanical Review* 51: 163–201.

Bakker, J. P. and F. Berendse. 1999. "Constraints in the Restoration of Ecological Diversity in Grassland and Heathland Communities." *Trends in Ecology and Evolution* 14: 63–68.

Belcher, J.W. and S.D. Wilson. 1989. "Leafy Spurge (*Euphorbia esula* L.) and the Species Composition of Mixed-grass Prairie." *Journal of Range Management* 42: 171–75.

Best, K.F., G.G. Bowes, A.G. Thomas and M.G. Maw. 1980. "The Biology of Canadian Weeds. 39. *Euphorbia esula.*" *Canadian Journal of Plant Science* 60: 651–63.

Burke, I.C., W.K. Lauenroth and D.P. Coffin. 1995. "Soil Organic Matter Recovery in Semiarid Grasslands: Implications for the Conservation Reserve Program." *Ecological Applications* 5: 793–801.

Burke, I.C., C.M. Yonker, W.J. Parton, C.V. Cole, K. Flach. 1989. "Texture, Climate, and Cultivation Effects on Soil Organic Matter Content in U.S. Grassland Soils." *Soil Science Society of America Journal* 53: 800–805.

Buyanovsky, G.A. and G.H. Wagner. 1998. "Carbon Cycling in Cultivated Land and Its Global Significance." *Global Change Biology* 4: 131–41.

Chapin, F.S. III, B.H. Walker, R.J. Hobbs, D.U. Hooper, J.H. Lawton, O.E. Sala, and D. Tilman. 1997. "Biotic Control Over the Functioning of Ecosystems." *Science* 277: 500–504.

Christian, J.M. and S.D. Wilson. 1999. "Long-term Impacts of an Introduced Grass in the Northern Great Plains." *Ecology* 80: 2397–2407.

Collins, S.L., A.K. Knapp, J.M. Briggs, J.M. Blair and E.M. Steinauer. 1998. "Modulation of Diversity by Grazing and Mowing in Native Tallgrass Prairie." *Science* 280: 745–47.

Costanza, R., R. d'Arge, R. de Groot, S. Farber, M. Grasso, B. Hannon, K. Limburg, S. Naeem, R.V. O'Neill, J. Paruelo, R.G. Raskin, P. Sutton and M. van den Belt. 1997. "The Value of the World's Ecosystem Services and Natural Capital." *Nature* 387: 253–60.

Coughenour, M.B. 1985. "Graminoid Responses to Grazing by Large Herbivores: Adaptations, Exaptation, and Interacting Processes." *Annals of the Missouri Botanical Garden* 72: 852–63.

Coupland, R.T. 1950. "Ecology of Mixed Prairie in Canada." *Ecological Monographs* 20: 271–315.

——. 1979. *Grassland Ecosystems of the World: Analysis of Grasslands and Their Uses.* International Biological Programme Report no. 18. London: Cambridge University Press.

Daily, G.C. (ed.). 1997. *Nature's Services: Societal Dependence on Natural Ecosystems.* Washington, D.C.: Island Press.

Daily, G.C., P.A. Matson and P.M. Vitousek. 1997. "Ecosystem Services Supplied by Soil." In Daily ed. *Nature's Services: Societal Dependence on Natural Ecosystems,* 113–32.

Davis, S.D. and D.C. Duncan. 1999. "Grassland Songbird Occurrence in Native and Crested Wheatgrass Pastures of Southern Saskatchewan." *Studies in Avian Biology* 19: 211–18.

Doebley, J.F. 1984. "'Seeds' of Wild Grasses: A Major Food of Southwestern Indians." *Economic Botany* 38: 52–64.

Dormaar, J.F., A. Johnston, and S. Smoliak. 1978. "Long-term Soil Changes Associated With Seeded Stands of Crested Wheatgrass in Southeastern Alberta, Canada." In *Proceedings of the First International Rangeland Congress,* 623–25.

Dormaar, J.F. and S. Smoliak. 1985. "Recovery of Vegetative Cover and Soil Organic Matter During Revegetation of Abandoned Farmland in a Semiarid Climate." *Journal of Range Management* 38: 487–91.

Dormaar, J.T., S. Smoliak and W.D. Willms. 1990. "Soil Chemical Properties During Succession from Abandoned Cropland to Native Range." *Journal of Range Management* 43: 260–65.

Dormaar, J.F., M.A. Naeth, W.D. Willms and D.S. Chanasyk. 1995. "Effect of Native Prairie, Crested Wheatgrass (*Agropyron cristatum* (L.) Gaertn.) and Russian Wildrye (*Elymus junceus* Fisch.) on Soil Chemical Properties." *Journal of Range Management* 48: 258–63.

Dyksterhuis, E.J. 1949. "Condition and Management of Rangeland Based on Quantitative Ecology." *Journal of Range Management* 2: 104–15.

——. 1958. "Ecological Principles in Range Evaluation." *Botanical Reviews* 24: 253–72.

Ehrenfeld, J.G. and L.A. Toth. 1997. "Restoration Ecology and the Ecosystem Perspective." *Restoration Ecology* 5: 307–17.

Environment Canada. 1993. *Canadian Climate Normals: 1961–1990. Volume 2. Prairie Provinces.* Ottawa: Environment Canada.

Everitt, J.H., C. J. DeLoach and W.G. Hart. 1989. "Using Remote Sensing for Detecting Brush and Weeds on Rangelands in the Southwestern United States." In E.S. Delfosse, ed., *Proceedings of the VII International Symposium for the Biocontrol of Weeds.* Italy: n.p., 585–93.

French, N. 1979. *Perspectives in Grassland Ecology.* New York: Springer-Verlag.

Friedel, M.H. 1991. "Range Condition Assessment and the Concept of Threseholds: A Viewpoint." *Journal of Range Management* 44: 422–26.

Gibson, D.J., T.R. Seastedt and J.M. Briggs. 1993. "Management Practices in Tallgrass Prairie: Large- and Small-scale Experimental Effects on Species Composition." *Journal of Applied Ecology* 30: 247–55.

Gilmore, M.R. 1977. *Uses of Plants by the Indians of the Missouri River Region.* Lincoln: University of Nebraska Press.

Gray, J.H. 1996. *Men Against the Desert.* Saskatoon: Western Producer Prairie Books.

Grilz, P.L. and J.T. Romo. 1995. "Management Considerations for Controlling Smooth Brome in Fescue Prairie." *Natural Areas Journal* 15: 148–56.

Hobbs, R.J. and L.F. Huenneke. 1992. "Disturbance, Diversity, and Invasion: Implications for Conservation." *Conservation Biology* 6: 324–37.

Holling, C.S. 1973. "Resilience and Stability of Ecological Systems." *Annual Review of Ecology and Systematics* 4: 1–23.

Holling, C.S. 1986. "The Resilience of Ecosystems: Local Surprise and Global Change." W.C. Clark and R.E. Munn, eds., *Sustainable Development of the Biosphere.* London: Cambridge University Press, 292–317.

Houghton, R.A., E.A. Davidson and G.M. Woodwell. 1998. "Missing Sinks, Feedbacks, and Understanding the Role of Terrestrial Ecosystems in the Global Carbon Balance." *Global Biogeochemical Cycles* 12: 25–34.

Huntley, N.J. 1991. "Herbivores and the Dynamics of Communities and Ecosystems." *Annual Review of Ecology and Systematics* 22: 477–503.

Jenny, H. 1980. *The Soil Resource.* New York: Springer-Verlag.

Joern, A. and K.H. Keeler. 1995. *The Changing Prairie: North American Grasslands.* New York: Oxford University Press.

Johnston, A. 1970. "A History of the Rangelands of Western Canada." *Journal of Range Management* 23: 3–8.

Kindscher, K. 1992. *Medicinal Wild Plants of the Prairie*. Lawrence: University Press of Kansas.

Kindscher, K. and L.L. Tieszen. 1998. "Floristic and Soil Organic Matter Changes After Thirty-five Years of Native Tallgrass Prairie Restoration." *Restoration Ecology* 6: 181–96.

Knapp, A.K. and T.R. Seastedt. 1986. "Detritus Accumulation Limits Productivity of Tallgrass Prairie." *BioScience* 36: 662–68.

Knapp, A.K., S.L. Collins, D.C. Hartnett and J.M. Blair. 1997. *Grassland Dynamics: Long-term Ecological Research in Tallgrass Prairie*. Lawrence: Kansas State University Press.

Knapp, A.K., J.M. Blair, J.M. Briggs, S.L. Collins, D.C. Hartnett, L.C. Johnson and E.G. Towne. 1999. "The Keystone Role of Bison in North American Tallgrass Prairie." *BioScience* 49: 39–50.

Lauver, C.L. 1997. "Mapping Species Diversity Patterns in the Kansas Shortgrass Region by Integrating Remote Sensing and Vegetation Analysis." *Journal of Vegetation Science* 8: 387–94.

Laycock, W.A. 1991. "Stable States and Thresholds of Range Condition on North American Rangelands: A Viewpoint." *Journal of Range Management* 44: 427–33.

Leach, M.K. and T.J. Givnish. 1996. "Ecological Determinants of Species Loss in Remnant Prairies." *Science* 273: 1555–58.

Leopold, A. 1949. *A Sand County Almanac*. London: Oxford University Press.

Lesica, P. 1995. "'Endless Sea of Grass'… No Longer." *Kelseya* 8: 8–9.

Lesica, P. and T.H. DeLuca. 1996. "Long-term Harmful Effects of Crested Wheatgrass on Great Plains Grassland Ecosystems." *Journal of Soil and Water Conservation* 51: 408–9.

Lym, R.G. and C.G. Messersmith. 1985. "Leafy Spurge Control With Herbicides in North Dakota: A 20-Year Summary." *Journal of Range Management* 38: 149–54.

Mack, R.N. 1981. "Invasion of *Bromus tectorum* L. "Into Western North America: An Ecological Chronicle." *Agro-Ecosystems* 7: 145–65.

——. 1989. "Temperate Grasslands Vulnerable to Plant Invasions: Characteristics and Consequences." In J.A. Drake, H.A. Mooney, F. di Castri, R.H. Groves, F.J. Kruger, M. Rejmánek, M. and M. Williamson, eds., *Biological Invasions: A Global Perspective*. Chichester, UK: John Wiley and Sons, Chichester, 155–79.

Manning, R. 1995. *Grassland: The History, Biology, Politics, and Promise of the American Prairie*. New York: Penguin Books.

Meyer, W.B. and B.L. Turner. 1992. "Human Population Growth and Global Land-use/Coverage Change." *Annual Review of Ecology and Systematics* 23: 39–62.

Morgan, J.P., D.R. Collicutt and J.D. Thompson. 1995. *Restoring Canada's Native Prairies: A Practical Manual*. Argyle, MB: Prairie Habitats.

Noy-Meir, I. and B.H. Walker. 1986. "Stability and Resilience in Rangelands." In P.J. Joss, P.W. Lynch and O.B. Williams, eds., *Rangelands: A Resource Under Siege*. London: Cambridge University Press, 21–25.

Olson, P.A. 1995. "Cultural Perception and Great Plains Grasslands." In A. Joern and K.H. Keeler, eds., *The Changing Prairie: North American Grasslands*. New York: Oxford University Press, 25–41.

PCAP Committee. 1998. *Saskatchewan Prairie Conservation Action Plan*. Regina: Canadian Plains Research Center.

Pearson, C.J. and R.L. Ison. 1997. *Agronomy of Grassland Ecosystems*. London: Cambridge University Press.

Peltzer, D.A., M.L. Bast, S.D. Wilson and A.K. Gerry. 2000. "Plant Diversity and Tree Responses Following Contrasting Disturbances in Boreal Forest." *Forest Ecology and Management* 127: 191–203.

Pepper, J. 1999. "Diversity and Community Assemblages of Ground-Dwelling Beetles and Spiders on Fragmented Grasslands of Southern Saskatchewan." M.Sc. thesis, University of Regina.

Peterson, G., C.R. Allen and C.S. Holling. 1998. "Ecological Resilience, Biodiversity, and Scale." *Ecosystems* 1: 6–18.

Pielou. E.C. 1991. *After the Ice Age: The Return of Life to Glaciated North America*. Chicago: University of Chicago Press.

Pimental, D., C. Harvey, P. Resosudarmo, K. Sinclair, D. Kurtz, M. McNair, S. Crist, L. Shpritz, L. Fitton, R. Saffouri and R. Blair. 1995. "Environmental and Economic Costs of Soil Erosion and Conservation Benefits." *Science* 267:1117–23.

Potyondi, B. 1995. *In Palliser's Triangle*. Saskatoon: Purich Publishing.

Rapport, D.J., R. Costanza and A.J. McMichael. 1998. "Assessing Ecosystem Health." *Trends in Ecology and Evolution* 13: 397–402.

Redente, E.F., M.E. Biondini and J.C. Moore. 1989. "Observations on Biomass Dynamics of a Crested Wheatgrass and Native Shortgrass Ecosystem in Southern Wyoming." *Journal of Range Management* 42: 113–18.

Reever Morghan, K.J. and T.R. Seastedt. 1999. "Effects of Soil Nitrogen Reduction on Nonnative Plants in Restored Grasslands." *Restoration Ecology* 7: 51–55.

Risser, P.G. 1988. "Abiotic Controls on Primary Productivity and Nutrient Cycles in North American Grasslands." In L.P. Pomeroy and J.J. Alberts, eds., *Concepts of Ecosystem Ecology: A Comparative View.* New York: Springer-Verlag.

Rogler, J.H. and R.L. Lorenz. 1983. "Crested Wheatgrass — Early History in the United States." *Journal of Range Management* 36: 91–93.

Romo, J.T., P.L. Grilz and L. Delanoy. 1994. "Selective Control of Crested Wheatgrass (*Agropyron cristatum* [L.] Gaertn. and *A. desertorum* Fisch.) in the Northern Great Plains." *Natural Areas Journal* 14: 308–9.

Rowe, J.S. 1969. "Lightning Fires in Saskatchewan Grassland." *Canadian Field Naturalist* 83: 317–24.

Sala, O.E. and J.M. Paruelo. 1997. "Ecosystem Services in Grasslands." In Daily ed. *Nature's Services: Societal Dependence on Natural Ecosystems,* 237–52.

Samson, F. and F. Knopf. 1994. "Prairie Conservation in North America." *BioScience* 44: 418–21.

——. 1996. Prairie conservation. Island Press, Washington, D.C.

Saskatchewan Environment and Resource Management (SERM). 1997. *The Prairie Ecozone: Our Agricultural Heritage.* Regina: SERM.

Schlesinger, W.H., J.F. Reynolds, G.L. Cunningham, L.F. Huenneke, W.M. Jarrell, R.A. Virginia and W.G. Whitford. 1990. "Biological Feedbacks in Global Desertification." *Science* 247: 1043–48.

Scott, G.A.J. 1995. *Canada's Vegetation: A World Perspective.* Montreal/Kingston: McGill-Queen's University Press.

Seastedt, T.R. and A.K. Knapp. 1993. "Consequences of Nonequilibrium Resource Availability Across Multiple Time Scales: The Transient Maxima Hypothesis." *American Naturalist* 141: 621–33.

Selleck, G.W., R.T. Coupland and C. Frankton. 1962. "Leafy Spurge in Saskatchewan." *Ecological Monographs* 32: 1–29.

Shantz, H.L. 1954. "The Place of Grasslands in the Earth's Cover of Vegetation." *Ecology* 35: 143–45.

Simberloff, D. and N. Gotelli. 1984. "Effects of Insularisation on Plant Species Richness in the Prairie-Forest Ecotone." *Biological Conservation* 29: 27–46.

Skold, M.D. 1989. "Cropland Replacement Policies and Their Effects on Land Use in the Great Plains." *Journal of Production Agriculture* 2: 197–201.

Spector, D. 1983. *Agriculture on the Prairies, 1870–1940.* Ottawa: Environment Canada.

Stanton, N.L. 1988. "The Underground in Grasslands." *Annual Review of Ecology and Systematics* 19: 573–89.

Statistics Canada. 2000. www.statscan.ca/english/pgdb/land (1996 census of agricultural land cover in Saskatchewan).

Stegner, W. 1962. *Wolf Willow: A History, a Story, and a Memory of the Last Plains Frontier.* New York: Viking Press.

Stein, B.A. and S.R. Flack. 1996. *America's Least Wanted: Alien Species Invasions of U.S. Ecosystems.* Arlington, VA: The Nature Conservancy.

Stewart, O.C. 1955. "Fire as the First Great Force Employed by Man." In W.L. Thomas, ed., *Man's Role in Changing the Face of the Earth.* Chicago: University of Chicago Press, 115–33.

Stohlgren, T.J., T.N. Chase, R.A. Pielke, T.G.F. Kittel and J. Baron. 1998. "Evidence that Local Land Use Practices Influence Regional Climate and Vegetation Patterns in Adjacent Natural Areas." *Global Change Biology* 4: 495–504.

Sutter, G.C. and R.M. Brigham. 1998. "Avifaunal and Habitat Changes Resulting from Conversion of Native Prairie to Crested Wheat Grass: Patterns at Songbird Community and Species Levels." *Canadian Journal of Zoology* 76: 869–975.

Tiessen, H., J.W.B. Stewart and J.R. Bettany. 1982. "Cultivation Effects on the Amounts and Concentration of Carbon, Nitrogen, and Phosphorous in Grassland Soils." *Agronomy Journal* 74: 831–35.

Tiessen, H., E. Cuevas and P. Chacon. 1994. "The Role of Soil Organic Matter in Sustaining Soil Fertility." *Nature* 371: 783–85.

Vallentine, J.F. 1989. *Range Management and Improvements.* New York: Academic Press.

Vogt, K.A., J.C. Gordon, J.P. Wargo, D.J. Vogt, H. Asbornsen, P.A. Palmiotto, H.J. Clark, J.L. O'Hara, W.S. Keeton, T. Patel-Weynand, E. Witten. 1997. *Ecosystems: Balancing Science with Management.* New York: Springer.

Walker, B.H. 1992. "Biodiversity and Ecosystem Redundancy." *Conservation Biology* 6: 18–23.

Walker, B. 1995. "Conserving Biological Diversity Through Ecosystem Resilience." *Conservation Biology* 9: 747–52.

Weaver, J.E. 1943. "Replacement of True Prairie by Mixed Prairie in Eastern Nebraska and Kansas." *Ecology* 24: 421–34.

——. 1954. *North American Prairie.* Lincoln, NE: Johnsen Publishing.

Weaver, J.E. and F.W. Albertson. 1956. *Grasslands of the Great Plains.* Lincoln, NE: Johnsen Publishing Company.

West, N.E. 1993. "Biodiversity of Rangelands." *Journal of Range Management* 46: 2–13.

Westman, W.A. 1977. "How Much are Nature's Services Worth?" *Science* 197: 960–64.

White, D.J., E. Haber and C. Keddy. 1993. *Invasive Plants of Natural Habitats in Canada: An Integrated Review of Wetland and Upland Species and Legislation Governing Their Control.* Ottawa: Environment Canada.

Wilson, S.D. and J.W. Belcher. 1989. "Plant and Bird Communities of Native Prairie and Introduced Eurasian Vegetation in Manitoba, Canada." *Conservation Biology* 3: 39–44.

Wilson, S.D. and A.K. Gerry. 1995. "Strategies for Mixed-grass Prairie Restoration: Herbicide, Tilling, and Nitrogen Manipulation." *Restoration Ecology* 3: 290–98.

Witkamp, M. 1971. "Soils as Components of Ecosystems." *Annual Review of Ecology and Systematics* 2: 83–110.

Wright, H.A. and A.W. Bailey. 1982. *Fire Ecology: United States and Southern Canada.* New York: John Wiley and Sons.

Chapter 5

Acton, D.A., G.A. Padbury, C.T. Stuchniff, L. Gallagher, D.A. Gauthier, L. Kelly, T.A. Radenbaugh, and J. Thorpe. 1998. *Ecoregions of Saskatchewan.* Regina: Canadian Plains Research Center and Saskatchewan Environment and Resource Management.

Austin, J.E. and M.R. Miller. 1995. "Northern Pintail (*Anas acuta*)." In A. Poole and F. Gill eds. *The Birds of North America,* No. 163. Washington, DC: The Academy of Natural Sciences, Philadelphia, and the American Ornithologists' Union.

Archibold, O.W. and M.R. Wilson. 1980. "The Natural Vegetation of Saskatchewan Prior to Agricultural Settlement." *Canadian Journal of Botany* 58: 2032–42.

Bechard, M.J. 1981. "Historic Nest Records of the Peregrine Falcon in Southern Saskatchewan and Southern Manitoba." *Blue Jay* 39: 182–83.

——. 1982. "Further Evidence for an Historic Population of Peregrine Falcons in Southern Saskatchewan." *Blue Jay* 40: 125.

Beauchamp, W.D., R.R. Koford, T.D. Nudds, R.G. Clark, and D.H. Johnson. 1996. "Long-term Declines in Nest Success of Prairie Ducks." *Journal of Wildlife Management* 60: 247–57.

Bancroft, J. 1993. "Observations on the Dove Family." *Blue Jay* 51: 150–52.

Bent, A.C. 1907. "Summer Birds of Southwestern Saskatchewan." *Auk* 24: 407–31.

——. 1908. "Summer Birds of Southwestern Saskatchewan." *Auk* 25: 25–35.

Bittner, R.A. 1988. "Bluebirds at Abernethy: History and 1988 Results." *Blue Jay* 46: 215–18.

Blue Jay. 1957. "More Starling Records." *Blue Jay* 15: 101–2.

Braun, C.E. 1998. "Sage Grouse Declines in Western North America: What are the Problems?" *Proc. West. Assoc. State Fish and Widl. Agencies* 78: 139–56.

Brazier, F. H. 1964a. "Status of the Mockingbird in the Northern Great Plains." *Blue Jay* 22: 63–75.

——. 1964b. "Additional Records of the Mockingbird." *Blue Jay* 22: 151.

Cadman, M.D. 1985. *Status Report on the Loggerhead Shrike* (Lanius ludovicianus) *in Canada.* Ottawa: Committee on the Status of Endangered Wildlife in Canada.

Cody, M.L. 1985. "Habitat Selection in Grassland and Open-country Birds." In M.L. Cody ed. *Habitat Selection in Birds.* Orlando, FL: Academic Press. 191–226.

Davis, S.K., D.C. Duncan, and M.A. Skeel. 1996. "The Baird's Sparrow: Status Resolved." *Blue Jay* 54: 185–91.

——. 1999. "Distribution and Habitat Associations of Three Endemic Grassland Songbirds in Southern Saskatchewan." *Wilson Bulletin* 111: 389–96.

Davis, S.K. and D.C. Duncan. 1999. "Grassland Songbird Occurrence in Native and Crested Wheatgrass Pastures of Southern Saskatchewan." *Studies in Avian Biology* 19: 211–18.

Dexter, J.S. 1922. "The European Gray Partridge in Saskatchewan." *Auk* 34: 253–54.

Downes, C.M., B.T. Collins, and B.P. McBride. 1999. *The Canadian Breeding Bird Survey 1966-1999.* Hull, QC: National Wildlife Research Centre, Canadian Wildlife Service.

Droege, S. and J.R. Sauer. 1994. "Are More North American Species Decreasing than Increasing?" In E.J.M. Hagemeiyer and T.J. Verstrael eds. *Bird Numbers 1992: Distribution, Monitoring and Ecological Aspects.* Voorburg/Haarlem: Statistics Netherlands. 97–306.

DuBowy, P.J. 1996. "Northern Shoveler (*Anas clypeata*)." In A. Poole and F. Gill eds. *The Birds of North America,* No. 217. Philadelphia/Washington, DC: The Academy of Natural Sciences and the American Ornithologists' Union.

Ecological Stratification Working Group (ESWG). 1995. *A National Framework of Canada.* Ottawa/Hull: Agriculture and Agri-Food Canada and Environment Canada.

Elliott, C. 1966. "An Extension of the Known Range of the Poorwill in Saskatchewan." *Blue Jay* 24: 78.

Epp, H.T. 1992. "Saskatchewan's Endangered Species and Spaces: Their Significance and Future. In P. Jonker ed. *Saskatchewan's Endangered Spaces: An Introduction.* Saskatoon: University of Saskatchewan. 47–108.

Erskine, A.J. 1979. "Man's Influence on Potential Nesting Sites and Populations of Swallows in Canada." *Canadian Field Naturalist* 93: 371–77.

Farley, F. L. 1925. "Changes in the Status of Certain Animals and Birds During the Past Fifty Years in Central Alberta." *Canadian Field Naturalist* 34: 200-202.

Fretwell, S. 1977. "Is the Dickcissel a Threatened Species?" *American Birds* 31: 923–32.

Furniss, O.C. 1944. "The European Starling in Central Saskatchewan." *Auk* 61: 469–70.

Fyfe, R.W. 1958. "Prairie Falcon Nesting Records in Saskatchewan." *Blue Jay* 14: 115–16.

Graul, W.D. and L.E.Webster. 1976. "Breeding Status of the Mountain Plover." *Condor* 78: 265–67.

Henderson, A.D. 1923. "The Return of the Magpie." *Oologist* 40: 142.

Houston, C.S. 1949. "The Birds of the Yorkton District." *Canadian Field Naturalist* 63: 215–41.

——. "Starling Records." *Blue Jay* 15: 9.

——. 1972. "The Passenger Pigeon in Saskatchewan." *Blue Jay* 30: 77–83.

——. 1977. "Changing Patterns of Corvidae on the Prairies." *Blue Jay* 35: 149–56.

——. 1979. "The Spread of the Western Kingbird Across the Prairies." *Blue Jay* 37: 149–57.

——. 1985. "Golden Eagles Nest Successfully in Trees." *Blue Jay* 43:131–33.

——. 1986. "Mourning Dove Numbers Explode on the Canadian Prairies." *American Birds* 40: 52–54.

——. 1999. "Decline in Upland Sandpiper Populations: History and Interpretations." *Blue Jay* 57: 136–42.

Houston, C.S. and M.J. Bechard. 1983. "Trees and the Redtailed Hawk in Southern Saskatchewan." *Blue Jay* 41: 99–109.

——. 1984. "Decline of the Ferruginous Hawk in Southern Saskatchewan." *American Birds* 38: 166–70.

Houston, C.S. and M.I. Houston. 1979. "Four Rancher-naturalists of the Cypress Hills, Saskatchewan." *Blue Jay* 37: 9–19.

——. 1997. "Saskatchewan Birds Species Which Increased With Settlement." *Blue Jay* 55: 90–96.

Houston, C.S. and A. Schmidt. 1981. "History of Richardson's Merlin in Saskatchewan." *Blue Jay* 39: 30–37.

Houston, C.S., D.G. Hjertaas, R.L. Scott, and P.C. James. 1996. "Experience With Burrowing Owl Nest-boxes in Saskatchewan, With Comment on Decreasing Range." *Blue Jay* 54: 136–40.

Houston, C.S. and J.K. Schmutz. 1999. "Changes in Bird Populations in Canadian Grasslands." *Studies in Avian Biology* 19: 87–94.

Igl, L.D. and D.H. Johnson. 1997. "Changes in Breeding Bird Populations in North Dakota: 1967 to 1992–93." *Auk* 114: 74–92.

Harrold, C.G. 1933. "Notes on the Birds Found at Lake Johnston and Last Mountain Lake, Saskatchewan, During April and May, 1922." *Wilson Bulletin* 45: 16–26.

Johnston, A. and S. Smoliak. 1976. "Settlements of the Grasslands and the Greater Prairie Chicken." *Blue Jay* 34: 153–56.

Hjertaas, D.G. 1979. "The Cattle Egret Arrives in Saskatchewan." *Blue Jay* 37, no. 2: 104–7.

Kalcounis, M.C., R.D. Csada and R.M. Brigham. 1991. "The Status and Distribution of the Common Poorwill in the Cypress Hills, Saskatchewan." *Blue Jay* 50: 38–44.

Knopf, F.L. 1994. "Avian Assemblages on Altered Grasslands." *Studies in Avian Biology* 15: 247–57.

——. 1995. "Declining Grassland Birds." In E.T. LaRoe, G.S Farris, C.E. Puckett, P.D. Doran, and M.J. Mac, eds. *Our Living Resources: A Report to the Nation on the Distribution, Abundance, and Health of U.S. Plants, Animals, and Ecosystems.* Washington, D.C.: U.S. Deptartment of the Interior. 296–98.

——. 1996. "Prairie Legacies—Birds." In: F.B. Samson and F.L. Knopf eds. *Prairie Conservation: Preserving North America's Most Endangered Ecosystem.* Washington, DC: Island Press. 135–48.

Knowles, E.H.M. 1938. "Polygamy in the Western Lark Sparrow." *Auk* 55: 675–76.

LeSchack, C.R., S.K. McKnight, and G.R. Hepp. 1997. "Gadwall (*Anas strepera*)." In A. Poole and F. Gill eds. *The Birds of North America*, No. 283. Philadelphia/Washington, DC: The Academy of Natural Sciences and the American Ornithologists' Union.

MacArthur, R.H. 1959. "On the Breeding Distribution Pattern of North American Migrants Birds." *Auk* 76: 318–25.

Macoun, J. 1900. *Catalogue of Canadian Birds.* Ottawa: Dawson Queen's Press.

McKim, L.T. 1926. "Arkansas Flycatcher Nesting at Melville, Sask." *Auk* 43: 370–71.

Mitchell, H.H. 1924. "Birds of Saskatchewan." *Canadian Field Naturalist* 38: 101–20.

Nero, R.W. 1993. "Lark Bunting — Western Prairie Marvel." *Blue Jay* 51: 30–33.

Nero, R.W., R.W. Lahrman, and R.G. Bard. 1958. "Dry-land Nest-site of a Western Grebe Colony." *Auk* 75: 347–49.

Owens, R.A. and M.T. Myres. 1973. "Effects of Agriculture Upon Populations of Native Passerine Birds of an Alberta Fescue Grassland." *Canadian Journal of Zoology* 51: 697–713.

Pearson, T.G. ed. 1936. *Birds of America.* New York: Doubleday and Co., Inc.

Potter, L.E. 1930. "Bird-life Changes in Twenty-five Years on Southwestern Saskatchewan." *Canadian Field Naturalist* 44: 147–49.

Price, J., S. Droege, and A. Price. 1995. *The Summer Atlas of North American Birds.* San Diego: Academic Press.

Radenbaugh, T.A. 1998. "Saskatchewan's Prairie Plant Assemblages: A Hierarchical Approach." *Prairie Forum* 23: 31–47.

Raine, W. 1892. *Bird-Nesting in North-West Canada.* Toronto: Hunter Rose.

Renaud, W.E. 1979a. "The Piping Plover in Saskatchewan: Status and Distribution." *Blue Jay* 37, no. 2: 91–103.

——. 1979b. "The Rock Wren in Saskatchewan: Status and Distribution." *Blue Jay* 37, no. 3: 139–48.

——. 1980. "The Long-Billed Curlew in Saskatchewan: Status and Distribution." *Blue Jay* 38, no. 4: 221–37.

Richardson, J. and W. Swainson. 1831. *Fauna Boreali Americana. Vol. 2. The Birds.* London: John Murray.

Risser, P.G., E.C. Birney, H.D. Blocker, S.W. May, W.J. Parton, and J.A. Wiens. 1981. *The True Prairie Ecosystem.* Stroudsburg, PA: Hutchinson Ross Publishing.

Rodney, K. 1982. "Cattle Egret Nesting Record for Saskatchewan." *Blue Jay* 40: 163–64.

——. 1993. "1991 White Pelican and Double Crested Cormorant Census in Saskatchewan." *Blue Jay* 51: 106–8.

Roy, J.F. n.d. *Birds of the Elbow.* Regina: Saskatchewan Natural History Society.

Sealy, S.G. 1971. "The Irregular Occurrences of the Dickcissel in Alberta, Manitoba and Saskatchewan." *Blue Jay* 29:12–16.

Selby, C.J. and M.J. Santry. 1996. *A National Ecological Framework for Canada: Data Model, Database and Programs.* Ottawa/Hull: Agriculture and Agri-Food Canada and Environment Canad.

Shadick, S.J. 1980. "Saskatchewan Breeding Records for Red Crossbill, Orchard Oriole, and Red-Headed Woodpecker." *Blue Jay* 38, no. 4: 247–49.

Shaw, W.T. 1944. "Extension of Breeding Range of the Western Burrowing Owl in Saskatchewan." *Auk* 61: 473–74.

Skaar, P.D. 1980. *Montana Bird Distribution — Mapping by Latilong.* Bozeman: n.p.

Skeel, M.A, and D. Hjertaas, 1993. "Saskatchewan Results of the 1991 International Piping Plover Census." *Blue Jay* 51: 36–46.

Smith, A.R. 1996. *Atlas of Saskatchewan Birds.* Regina: Saskatchewan Natural History Society.

Soper, J.D. 1970. *Unpublished Field Notes on the Birds Observed and Collected in the Province of Saskatchewan, Canada, in 1914, 1921, 1927, and from July 1937 to September 1947.* Edmonton: University of Alberta.

Stewart, C. 1951. "European Starling." *Blue Jay* 9: 9

Stewart, R.E. 1975. *Breeding Birds of North Dakota.* Fargo: Tri-College Center for Environmental Studies.

Stewart, R.E. and H.A. Kanturd. 1972. "Population Estimates of Breeding Birds in North Dakota." *Auk* 89: 766–88.

Sutter, G.C., T. Troupe and M. Forbes. 1995. "Abundance of Baird's Sparrows, *Ammodramus bairdii*, in Native Prairie and Introduced Vegetation." *Ecoscience* 2: 344–48.

Sutter, G.C., S.K. Davis, and D.C. Duncan. 2000. "Grassland Songbird Abundance Along Roads and Trails in Southern Saskatchewan." *Journal of Field Ornithology* 71: 110–16.

Todd, W.E.C. 1947. "Notes on the Birds of Southern Saskatchewan." *Annals of the Carnegie Museum* 30: 383–421.

Wedgwood, J.A. 1976. "Burrowing Owls in South-central Saskatchewan." *Blue Jay* 34: 26–44.

——. 1992. "Common Nighthawks in Saskatoon." *Blue Jay* 50: 211–17.

Weins, J.A. 1974. "Climate Instability and the 'Ecological Saturation' of Bird Communities in North American Grasslands." *Condor* 74: 385–400.

——. 1989. *The Ecology of Bird Communities: Volume 1. Foundations and Patterns.* Cambridge: Cambridge University Press.

Welicome, T.L. and E.A. Haug. 1995. *Second Update of Status Report on the Burrowing Owl, Speotypto canicularia, in Canada.* Ottawa: Committee on the Status of Endangered Wildlife in Canada.

Williams, M.Y. 1946. "Notes on the Vertebrates of the Southern Plains of Canada, 1923–1926." *Canadian Field Naturalist* 60: 47–60.

Wilson, M.F. 1976. "The Breeding Distribution of North American Migrant Birds: A Critique of MacArthur (1959)." *Wilson Bulletin* 88: 582–87.

Chapter 6

Barney, D. 1995. "Pushbutton Democracy: The Reform Party and the Real World of Teledemocracy." Paper presented to the Annual Meeting of the Canadian Political Science Association, Montreal, June 4-6.

Bell, E. 1993. *Social Classes and Social Credit*. Montreal and Kingston: McGill-Queen's University Press.

Berton, Pierre. 1991. *The Great Depression, 1929–1939*. Toronto: Penguin Books.

Canovan, M. 1981. *Populism*. New York: Harcourt Brace Jovanovich.

Clark, S.D., J.P. Grayson and L.M. Grayson. 1975. "General Introduction. The Nature of Social Movements." In Clark, S.D., J.P. Grayson, and L.M. Grayson eds. *Prophecy and Protest: Social Movements in Twentieth-Century Canada*. Toronto: Gage Educational Publishing Ltd.

Conway, J.F. 1978. "Populism in the United States, Russia, and Canada: Explaining the Roots of Canada's Third Parties," *Canadian Journal of Political Science* 11, no. 1: 99–124.

——. 1994. *The West. The History of a Region in Confederation*. Toronto: Lorimer.

Dabbs, F. 1995. *Ralph Klein: A Maverick Life*. Vancouver: Greystone Books.

Dobbin, M. 1991. *Preston Manning and the Reform Party*. Toronto: Lorimer and Company Pub.

Epp, R. 2000. "Tory MLAs Face a New Choice: Team Players or Voter Supporters." *Edmonton Journal*, 22 April, A15.

Finkel, A. 1989. *The Social Credit Phenomenon in Alberta*. Toronto: University of Toronto Press.

Flanagan, T. 1995. *Waiting for the Wave: The Reform Party and Preston Manning*. Toronto: Stoddart.

Gibbins, R. 1980. *Prairie Politics and Society: Regionalism in Decline*. Toronto: Butterworths.

Gibbins, R., and S. Arrison. 1995. *Western Visions. Perspectives on the West in Canada*. Peterborough: Broadview Press.

Harrison, T. 1995a. *Of Passionate Intensity: Right-Wing Populism and the Reform Party of Canada*. Toronto: University of Toronto Press.

——. 1995b. "Making the Trains Run on Time: Corporatism in Alberta." In G. Laxer and T. Harrison eds. *The Trojan Horse: Alberta and the Future of Canada*. Montreal: Black Rose Books.

Harrison, T., and J. Kachur, eds. 1999. *Contested Classrooms. Education, Globalization, and Democracy in Alberta*. Edmonton: University of Alberta Press and Parkland Institute.

Harrison, T., W. Johnston and H. Krahn. 1996. "Special Interests and/or New Right Economics? The Ideological Bases of Reform Party Support in Alberta in the 1993 Federal Election." *Canadian Review of Sociology and Anthropology* 33, no. 2: 159–79.

Harrison, T., and G. Laxer. 1995. "Introduction." In Laxer and Harrison, *The Trojan Horse*.

Irving, J. 1959. *The Social Credit Movement in Alberta*. Toronto: University of Toronto Press.

Kachur, J. 1999. "Orchestrating Delusions: Ideology and Consent in Alberta." In Harrison and Kachur, *Contested Classrooms*.

Kachur, J., and T. Harrison. 1999. "Introduction." In Harrison and Kachur, *Contested Classrooms*.

Laclau, E. 1977. *Politics and Ideology in Marxist Theory: Capitalism — Fascism — Populism*. London: New Left Books.

Laxer, G. and T. Harrison eds. 1995. *The Trojan Horse: Alberta and Future of Canada*. Montreal: Black Rose Books.

Laycock, D. 1990. *Populism and Democratic Thought in the Canadian Prairies, 1910 to 1945*. Toronto: University of Toronto Press.

——. 1994. "Reforming Canadian Democracy? Institutions and Ideology in the Reform Party Project." *Canadian Journal of Political Science* 28, no. 2: 213–48.

Lipset, S.M. 1968. *Agrarian Socialism: The Cooperative Commonwealth Federation in Saskatchewan*. Garden City, New York: Doubleday.

Lisac, M. 1995. *The Klein Revolution*. Edmonton: NeWest Press.

Macpherson, C.B. 1953. *Democracy in Alberta: Social Credit and the Party System*. Toronto: University of Toronto Press.

Mallory, J.R. 1954. *Social Credit and the Federal Power in Canada*. Toronto: University of Toronto Press.

Manning, P. 1992. *The New Canada*. Toronto: Macmillan Canada.

Monto, T. 1989. *The United Farmers of Alberta — A Movement, A Government*. Edmonton: Crang Publishing.

Morton, W. 1950. *The Progressive Party in Canada*. Toronto: University of Toronto Press.

Patten, S. 1993. "Populist Politics? A Critical Re-examination of 'Populism' and the Character of the Reform Party's Populist Politics." Paper presented to the Annual Meeting of the Canadian Political Science Association, Ottawa, June 6-8.

Richards, J. 1981. "Populism: A Qualified Defence." *Studies in Political Economy* 5: 5–27.

Richards, J., and L. Pratt. 1979. *Prairie Capitalism. Power and Influence in the New West*. Toronto: McClelland and Stewart.

Schmitter, P.C. 1993. "Corporatism." In J. Krieger ed. *The Oxford Companion to Politics of the World.* Oxford: Oxford University Press.

Sharpe, S., and D. Braid. 1992. *Storming Babylon. Preston Manning and the Rise of the Reform Party.* Toronto: Key Porter Books.

Sinclair, P. 1979. "Class Structure and Populist Protest: The Case of Western Canada." In C. Caldarola ed. *Society and Politics in Alberta: Research Papers.* Toronto: Methuen.

Stewart, D. 1995. "Klein's Makeover of the Alberta Conservatives." In Laxer and Harrison, *The Trojan Horse.*

Stingel, J. 2000. *Social Discredit: Anti-Semitism, Social Credit, and the Jewish Response.* Montreal: McGill-Queen's University Press.

Taft, K. 1997. *Shredding the Public Interest. Ralph Klein and 25 Years of One-Party Government.* Edmonton: University of Alberta Press and Parkland Institute.

Taft, K., and G. Steward. 2000. *Clear Answers. The Economics and Politics of For-Profit Medicine.* Edmonton: Duval Publishing, Parkland Institute, and University of Alberta Press.

Zakuta, L. 1975. "Membership in a Becalmed Social Movement." In Clark et al., *Prophecy and Protest.*

Chapter 7

Barger, Harold, and Hans Landsberg. 1942. *American Agriculture, 1899–1939: A Study of Output, Employment and Productivity.* New York: National Bureau of Economic Research.

Berry, Brian J.L. et al. 1988. *Market Centers and Retail Location.* Englewood Cliffs, NJ: Prentice Hall.

Britnell, G.E. 1939. *The Wheat Economy.* Toronto: University of Toronto Press.

Fowke, V.C. 1957. *The National Policy and The Wheat Economy.* Toronto: University of Toronto Press.

Furtan, W.H., and G.E. Lee. 1977. "Economic Development of the Saskatchewan Wheat Economy." *Canadian Journal of Agricultural Economics* 25: 15–28.

Hodge, Gerald. 1965. The Prediction of Trade Centre Viability in the Great Plains. *Papers, Regional Science Association* 15: 87–115.

Kouri, D. M. 1999. "Health Care Regionalization in Saskatchewan: An Exercise in Democracy." M.A. Thesis, University of Regina, Regina, Saskatchewan.

Mackintosh, W.A. 1934. *Prairie Settlement.* Toronto: MacMillan.

Martin, Chester, and Arthur S. Morton. 1938. *History of Prairie Settlement and 'Dominion Lands' Policy.* Toronto: MacMillan.

Olfert, M. Rose and Jack C. Stabler, 1994. "Industrial Restructuring of the Prairie Labour Force: Spatial and Gender Impacts." *Canadian Journal of Regional Science* 17: 133–52.

——. 1998. "Spatial Dimensions of Rural, Gender Specific Labour Force Commuting Patterns." *Australasian Journal of Regional Studies* 4: 253–74.

——. 1999. "Multipliers in a Central Place Hierarchy." *Growth and Change* 30: 288–302.

Phillips, W.G. 1956. *The Agricultural Implements Industry in Canada.* Toronto: University of Toronto Press.

Saskatchewan Economic Development. 1996. *REDAs.* <<http://www.gov.sk.ca/econdev/starting/reda.shtml.>> Regina: Economic and Cooperative Development.

Saskatchewan Health. 1998. *Covered Population.* Regina: Queen's Printer.

Saskatchewan Municipal Government. 1993. Partnerships for Municipal Services Delivery: A Review of Intermunicipal Arrangements in Saskatchewan. Discussion Paper, in *Report of the Minister's Advisory Committee on Inter-Community Co-operation and Community Quality of Life.* Regina: Saskatchewan Municipal Government.

Stabler, Jack C. 1996. Economics and Multicommunity Partnerships. *Canadian Journal of Regional Science* 19, no. 1: 71–94.

Stabler, Jack C. and M. Rose Olfert. 1992. *Restructuring Rural Saskatchewan: The Challenge of the 1990s.* Regina: University of Regina, Canadian Plains Research Center.

——. 1994. Farm Structure and Community Viability in the Northern Great Plains. *Review of Regional Studies,* January.

——. 1996. *Rural Communities in an Urbanizing World: An Update to 1995.* Regina: University of Regina, Canadian Plains Research Center.

——. 1999. "Impact of the Termination of the Crow Subsidy in the Context of the Economic Restructuring in the East Central Region." Working paper, Department of Agricultural Economics, University of Saskatchewan, Saskatoon, Saskatchewan.

——. 2000. "Functional Economic Areas: A Framework for Municipal Restructuring." Working Paper, Department of Agricultural Economics. Saskatoon: University of Saskatchewan.

Stabler, Jack C., M. Rose Olfert and Jonathan B. Greuel. 1996. "Spatial Labour Markets and the Rural Labour Force." *Growth and Change* 27: 206–30.

Wensley, Mitch R.D. and Jack C. Stabler. 1998. "Demand Threshold Estimation for Business Activities in Rural Saskatchewan." *Journal of Regional Science* 38 no. 1: 155–77.

Chapter 8

Braband, Lynn. 1986. "Railroad Grasslands as Bird and Mammal Habitats in Central Iowa." In Gary K. Clambey and Richard H. Pemble eds. *The Prairie: Past, Present, and Future: Proceedings of the Ninth North American Prairie Conference, July 29 to August 1, 1984.* Fargo: Tri-College University Center for Environmental Studies.

Goodman, Lowell and Leo Reinhold. 1993. "Historical Geography: World War I and 36.6 Rail Miles of History." *Bulletin, Association of North Dakota Geographers* 43: 25–33.

Granata, D.S. and K.L. McGown. 1997. "Analysis and Transfer of Commodity-Specific Shortlines in Western Canada — Case Study: CN Avonlea Subdivision." In *Transportation: Emerging Realities.*

Leopard, John. 1999. "Twilight of the GMDI." *Railfan & Railroad* 18, no. 7: 36–41.

Lewis, E.A. 1996. *American Shortline Railway Guide.* Waukesha, WI: Kalmbach Publishing.

McDonnell, Greg 1997. *Passing Trains.* North York, ON: Stoddart.

——. 1998. *Wheat Kings.* Erin, ON: Boston Mills Press.

Paul, Alec H. 1997. "Shortlines, Mainlines, Branchlines, Dead Lines: Rural Railways in Southwestern Saskatchewan in the 1990s." In John Welsted and John Everitt eds. *The Yorkton Papers: Research by Prairie Geographers.* Brandon, MB: Brandon University. 122–35.

Paul, Russell and Georgia Bratvold. 1992. "Grain Transport: Is It Profitable or Not?" *The Leader-Post.* 19 May.

Sorenson, L.O. 1984. "Some Impacts of Rail Regulatory Changes on Grain Industries." *American Journal of Agricultural Economics* 66: 645–50.

Williams, A. and J. Everitt 1989. "An Analysis of Settlement Development in Southwest Manitoba: The Lenore Extension 1902-1982." In H.J. Selwood and J.C. Lehr eds. *Prairie and Northern Perspectives: Geographical Essays.* Winnipeg: University of Winnipeg: 87–105.

Chapter 9

Adams, F.G. 1873. *The Homestead Guide, Describing the Great Homestead Region in Kansas and Nebraska and containing the Homestead, Pre-emption and Timber Bounty Laws, and a Map of the Country Described.* Waterville: F.G. Adams.

Blouet, B. and M. Lawson eds. 1975. *Images of the Plains.* Lincoln: University of Nebraska Press.

Bone, N. 1991. *The Aurora-Sun-Earth Interactions.* Chichester, England: Ellis Horwood.

Bowden, M. 1976. "The Great American Desert in the American Mind: The Historiography of a Geographical Notion" in D. Lowenthal and M.Bowden eds. *Geographies of the Mind.* New York: Oxford University Press.

Brown, R. 1948. *Historical Geography of the United States.* New York: Harcourt, Brace, and World, Inc.

Copley, J. 1867. *Kansas and the Country Beyond, on the line of the Union Pacific Railway Division, from the Missouri to the Pacific Ocean, partly from Personal Observation, and partly from information drawn from authentic sources.* Philadelphia: J B. Lippincott and Co.

Changnon, S.A., Jr., 1977. "The Climatology of Hail in North America." InG.B. Foote and C.A. Knight eds. *Hail: A Review of Hail Science and Hail Suppression,* 107–28. Boston: American Meteorological Society.

Dary, David. 1987. *More True Tales of Old Time Kansas.* Lawrence: University of Kansas.

Davis, K. 1977. *Kansas, A Bicentennial History.* New York: Norton.

Eagleman, J.R., and J.E. Simmons. 1985. "Weather." in J.T. Collins ed. *Natural Kansas.* Lawrence: University Press of Kansas.

Emmons, D.M. 1975. "The Influence of Ideology on Changing Environmental Images." In B.W. Blouet and M. Lawson eds. *Images of the Plains, the Role of Human Nature in Settlement.* Lincoln: University of Nebraska Press.

Fleming. J.R. 1990. *Meteorology in America, 1800–1870.* Baltimore: The Johns Hopkins University Press.

Flora, S.D. 1948. "Preface, Report of the Kansas State Board of Agriculture." *Climate of Kansas* 67, no. 285: iv-v.

Goodnow, Isaac. 1863. Diary (unpublished manuscript in the special collections of the Riley County Historical Society, Manhattan, Kansas).

Greeley, H. 1860. *An Overland Journey from New York to San Francisco in the Summer of 1859.* New York: C.M. Saxton, Barker & Co.

Henry, A. 1894. "Early Individual Observers in the United States in the US Department of Agriculture Weather Bureau." *Report of the International Meteorological Congress.* Washington DC: Weather Bureau.

Henry, J. 1886. *Scientific Writings of Joseph Henry, Vol. 1.* Washington D.C.: The Smithsonian Institution.

Kansas as She is: the Greatest Fruit, Stock and Grain Country in the World. 1870. Lawrence: Kansas Publishing Company.

Lawson, M.P. 1974. "The Climate of the Great American Desert." *Reconstruction of the Climate of Western Interior U.S. 1800–1850.* Lincoln: University of Nebraska Press.

Malin, J.C. 1990. "Some Reflections on Cultural Inheritance and Originality" in P. Stuewe, ed., *Kansas Revisited: Historical Images and Perspective.* Lawrence: University of Kansas Press.

Manley, R. 1993. Land and Water in 19th Century Nebraska "The Desert Shall Rejoice and Blossom as the Rose!." In C.A. Flowerday ed. *Flat Water: a History of Nebraska and Its Water.* Lincoln: University of Nebraska.

McAllaster, B. n.d. *Kansas: The Golden Belt Lands Along the Line of the Kansas Division of the U.P.R.Y.* Kansas City, MO: n.p.

Meinig, D. 1993. *The Shaping of America, a Geographical Perspective on 500 Years of History: Continental America, 1800–1867, Volume 2.* New Haven: Yale University Press.

Mohler, C. 1948. *Climate of Kansas, Report of the Kansas State Board of Agriculture.* Topeka: Ferd. Voiland, State Printer.

National Weather Service Archives, 1858–1873.

Olson, P. A. 1995. "Cultural Perception and the Great Plains Grasslands" in A. Joern and K. Keeler, eds., *The Changing Prairie, North American Grasslands.* New York: Oxford University Press.

Rhees, W.J. 1859. *An Account of the Smithsonian Institution, Its Founder, Building, Operations, etc.* Washington: Thomas McGill, Printer.

Rosenberg, N. 1986. "Climate of the Great Plains Region of the United States." *Great Plains Quarterly* 7: 22–32.

Self, H. 1978. *Environment and Man in Kansas: A Geographical Analysis.* Lawrence: The Regents Press of Kansas.

Snow, F. 1873. "Climate of Kansas." *Transactions of the Kansas Academy of Science* 5: 15–18.

Travis, P.D. 1978. "Changing Climate in Kansas: A Late Nineteenth Century Myth." *Kansas History* 1, no. 1: 48–58.

Webb, T.H. 1856. *Information for Kansas Immigrants.* A. Mudge and Son.

Wilder, D.W. 1886. *Annals of Kansas.* Topeka: Kansas Publishing House.

Willard, J.T. 1940. *History of the Kansas State College of Agriculture and Applied Science.* Manhattan: Kansas State College Press.

Zornow, W.T. 1957. *Kansas: A History of the Jayhawk State.* Norman: University of Oklahoma Press.

Chapter 10

Acton, D.F. and L.J. Gregorich. 1995. *The Health of Our Soils — Toward Sustainable Agriculture in Canada.* Ottawa: Agriculture and Agri-Food Canada.

Agriculture and Agri-Food Canada. 1999. "Medium Term Policy Baseline." Ottawa: <<http://www. agr.ca/policy/epad/english/pubs/mtb/mtbindex.htm>>

——. 2000. Climate Change Table. Options Report: "Reducing Greenhouse Gas Emissions from Canadian Agriculture."

Anderson, D.W. 1995. "Decomposition of Organic Matter and Carbon Emissions from Soils." In R. Lal, J. Kimlbe, E. Levine, and B.A. Stewart eds. *Advances in Soil Science: Soils and Global Change.* Boca Raton, FL: Lewis.

Beare, M.H., P.F. Hendrix and D.C. Coleman. 1994b. "Water-stable Aggregates and Organic Matter Fractions in Conventional and No-tillage Soils." *Soil Sci. Soc. Am. J.* 58: 777–86.

Bonneau, M., Kulshreshtha, S., M. Boehm, and J. Giraldez. 1999. *Canadian Economic and Emissions Model for Agriculture (CEEMA Version 1.0). Report 3. Preliminary Results of Selected Scenarios, Technical Tables.* Ottawa: Agriculture and Agri-Food Canada.

Bruce, J.P., M. Frome, E. Haites, H. Janzen, R. Lal, and K. Paustian. 1999. "Carbon Sequestration in Soils." *J. Soil and Water Conservation.* 52: 382–89.

Bruce, R.R., G.W. Langdale and A.L. Dillard. 1990. "Tillage and Crop Rotation Effect on Characteristics of a Sandy Surface Soil." *Soil Sci. Soc. Am. J.* 54: 1744–47.

Campbell, C.A. and R.P. Zentner. 1993. "Soil Organic matter as Influenced by Crop Rotations and Fertilization." *Soil Sci. Soc. Am. J.* 57: 1034–40.

Campbell, C.A., G.P. Lafond, A.J. Leyshon, R.P. Zentner and H.H. Janzen. 1991. "Effect of Cropping Practices on the Initial Potential Rate of N Mineralization in a Thin Black Chernozem." *Can. J. Soil Sci.* 71: 43–53.

Carefoot, J.M., M. Nyborg and C.W. Lindwall. 1990. "Tillage-induced Soil Changes and Related Grain Yield in a Semi-arid Region." *Can J. Soil. Sci.* 70: 203–14.

Carter, M.R. 1992. "Characterizing the Soil Physical Condition in Reduced Tillage Systems for Winter Wheat on a Fine Sandy Loam Using Small Cores." *Can. J. Soil Sci.* 72: 395–402.

Doran, J.W. and D.M. Linn. 1994. "Microbial Ecology of Conservation Management Systems." In J.L.

Hatfield and B.A. Stewart eds. *Soil Biology: Effects on Soil Quality. Advances in Soil Science*. Boca Raton, FL: Lewis.

Doran, J.W. and T.B. Parkin. 1994. "Defining and Assessing Soil Quality." In J.W. Doran, D.S. Coleman, D.F. Bezdicek and B.A. Stewart, eds., *Defining Soil Quality for a Sustainable Environment*. Madison, WI: Soil Sci. Soc. Am. Inc.

Environment Canada. 1997. *Canada's Second National Report on Climate Change*. Ottawa: Environment Canada.

Gallaher, R.N. and M.B. Ferrer. 1987. "Effect of No-tillage Versus Conventional Tillage on Soil Organic Matter and Nitrogen Contents." *Commun. in Soil Sci. Plant Anal.* 18: 1061–76.

Gregorich, E.G., M.R. Carter, D.A. Angers, C.M. Monreal and B.H. Ellert. 1994. "Towards a Minimum Data Set to Assess Soil Organic Matter Quality in Agricultural Soils." *Can. J. Soil Sci.* 74: 367–86.

Halvorson, A.D., C. Reule, and R. Follett. 1999. "Nitrogen Fertilization Effects on Soil Carbon and Nitrogen in a Dryland Cropping System." *Soil Sci. Soc. Am. J.* 63: 912–17.

Hamilton, K. 1993. *Energy Consumption, Environmental Perspectives 1993*. Ottawa: Statistics Canada.

Havlin, J.L., D.E. Kissel, L.D. Maddux, M.M. Claassen and J.H. Long. 1990. "Crop Rotation and Tillage Effects on Soil Organic Carbon and Nitrogen." *Soil Sci. Soc. Am. J.* 54: 448–52.

Hendrix, P.F., R.W. Parmelee, D.A. Crossley, D.C. Coleman, E.P. Odum and P.M. Groffman. 1986. "Detritus Food Webs in Conventional and No-tillage Agroecosystems." *BioScience* 36: 374–80.

Horner, G.L., J. Corman, R.E. Howitt, C.A. Carter, and R.J. MacGregor. 1992. *The Canadian Regional Agricultural Model: Structure, Operations and Development*. Ottawa: Agriculture and Agri-Food Canada.

Houghton, J.T., L.G. Meira Filho, B. Lim, K Treanton, I. Mamaty, U. Bonduki, D.J. Griggs and B.A. Callender eds. 1997. *Revised 1996 IPCC Guidelines for National Greenhouse Gas Inventories*, Volume 3: *Greenhouse Gas Inventory Reference Manual*. United Kingdom: IPCC/OECD/IEA.

Janzen, H.H., R. L. Desjardins, R. Asselin and B. Grace eds. 1998. *The Health of Our Air: Towards Sustainable Agriculture in Canada*. Ottawa: Agriculture and Agri-Food Canada.

Janzen, H.H. 1987. "Soil Organic Matter Characteristics After Long-term Cropping to Various Spring Wheat Rotations." *Can. J. Soil Sci.* 67: 845–56.

Junkins, B., S.N. Kulshreshtha, R. MacGregor, R. Gill, C. Dauncey, R. Desjardins, M. Boehm, P. Thomassin, A. Weersink, K Parton, and J. Cleary. 2000. Analyses of Strategies for Reducing Greenhouse Gas Emissions from Canadian Agriculture: Technical Report to the Agriculture and Agri-Food Table. Ottawa: Agriculture and Agri-Food Canada.

Kulshreshtha, S.N., B. Junkins and R.L. Desjardins. 1999. "Regional Pattern of Greenhouse Gas Emissions from the Agriculture and Agri-Food Sector in Canada." Paper presented at the 49th North American Meetings of the Regional Sciences International, November 11-14, 1999. Montreal.

Kulshreshtha, S.N., M. Bonneau, M. Boehm, and J. Giraldez. 1999. *Canadian Economic and Emissions Model for Agriculture (CEEMA Version 1.0). Report 1. Model Description*. Ottawa: Agriculture and Agri-Food Canada

Lal, R., A.A. Mahboubi and N.R. Fausey. 1994. "Long-term Tillage and Rotation Effects on Properties of a Central Ohio Soil." *Soil Sci. Soc. Am. J.* 58: 517–22.

Larney, F.J., C.W. Lindwall, R.C. Izaurralde and A.P. Moulin. 1994. "Tillage Systems for Soil and Water Conservation on the Canadian Prairie." In M.R. Carter ed. *Conservation Tillage in Temperate Agroecosystems*. Boca Raton, FL: Lewis.

Mahboubi, A.A., R. Lal and N.R. Faussey. 1993. "Twenty-eight Years of Tillage Effects on Two Soils in Ohio." *Soil Sci. Soc. Am. J.* 57: 506–12.

Martin, R., and A. Fredeen. 1999. *Effect of Management of Grasslands on Greenhouse Gas Balance*. Ottawa: Agriculture and Agri-Food Canada

McConkey, B.G., B.C. Liang, and C.A. Campbell. 1999. *Estimating Gains of Soil Carbon Over a 15-year Period Due to Changes in Fallow Frequency, Tillage System, and Fertilization Practices for the Canadian Prairies (An Expert Opinion)*. Ottawa: Agriculture and Agri-Food Canada.

Neitzert, F., K. Olsen and P. Collas. 1999. *Canada's Greenhouse Gas Inventory, 1997 Emissions and Removals with Trends*. Ottawa: Environment Canada.

Parton, W. J., S.D.S. Schimel, C.V. Cole and D.S. Ojima. 1987. "Analysis of Factors Controlling Soil Organic Matter Levels in the Great Plains Grasslands." *Soil Science Society of America Journal* 51: 1173–79.

Smith, W., R.L. Desjardins and B. Grant. 2000. "Estimated Changes in Soil Carbon Associated With Agricultural Practices in Canada." *Can. J. Soil Sci.* (submitted).

Statistics Canada. 1977. *Industrial Consumption of Energy Survey*. Ottawa: Statistics Canada.

Watson, R.T., H. Rodhe, H. Oeschger, U. Siegenthaler. 1990. "Greenhouse Gases and Aerosols." In J.T. Houghton, G.J. Jenkins and J.J. Ephraums, eds., *Climate Change: The IPCC Scientific Assessment*. Cambridge: Cambridge University Press.

Wood, C.W. and J.H. Edwards. 1992. "Agroecosystem Management Effects on Soil Carbon and Nitrogen." *Agric., Ecosystems, and Env.* 39: 123–38.